Sikhar Patranabis · Debdeep Mukhopadhyay
Editors

Fault Tolerant Architectures for Cryptography and Hardware Security

Springer

Editors
Sikhar Patranabis
Department of Computer Science and Engineering
Indian Institute of Technology Kharagpur
Kharagpur, West Bengal
India

Debdeep Mukhopadhyay
Department of Computer Science and Engineering
Indian Institute of Technology Kharagpur
Kharagpur, West Bengal
India

ISSN 2367-3478 ISSN 2367-3486 (electronic)
Computer Architecture and Design Methodologies
ISBN 978-981-10-1386-7 ISBN 978-981-10-1387-4 (eBook)
https://doi.org/10.1007/978-981-10-1387-4

Library of Congress Control Number: 2018932188

Printed on acid-free paper

This Springer imprint is published by Springer Nature
The registered company is Springer Nature Singapore Pte Ltd.
The registered company address is: 152 Beach Road, #21-01/04 Gateway East, Singapore 189721, Singapore

Preface

When a secret is revealed, it is the fault of the man who confided it.

Faults can be catastrophic for cryptosystems! In most cases, even a single well-formed fault is sufficient to reveal secret keys of the underlying ciphers. Even mathematically strong ciphers like the Advanced Encryption Standard (AES), RSA, are all vulnerable against such *fault attacks*. On the other hand, to meet the real-time requirements cryptosystems are often implemented in hardware platforms (in the form of FPGAs, ASICs), and as highly optimized software libraries (like OpenSSL) to be executed on a wide range of processors. Reliability of such complex designs, both on hardware and software is a serious issue. The problem becomes even more challenging than a standard reliability problem due to the fact that the reliability issue does not only lead to a failure, but could lead to a complete collapse of the cryptosystems. Like all security problems, here also there are two entities: the fault attacker and a fault attack-resistant designer. The former tries to develop novel fault injection mechanisms, fault analysis techniques which impose less restrictions on the fault injection and are based on more practical and achievable fault models. On the other hand, the designers' role is to evaluate the applicability of classical fault tolerance techniques to mitigate these threats, and to augment the defenses by dedicated methodologies. The designers also need a thorough understanding of the fault models, and the exploitable fault space to develop safeguards, which are sufficient to thwart the attacks. For making the treatment complete, it is also necessary to understand the reliability issues in modern day processors, to comprehend the threats in triggering these menacing attacks. Finally, it is also desirable to develop automated tools to assist in the fault analysis process to unearth new fault attacks against the cryptosystems. This would indeed reduce design cycles and help designers in the long run to develop fault-resistant systems with lesser effort.

The book tries to cover all these aspects and present the reader with a one-stop platform to develop comprehensive knowledge in this research area. A brief topic wise summary of the book is provided underneath to help the reader foresee his journey through the book:

- **Fault Analysis Methods and Fault Models**: There are different types of fault analysis methods which have been developed, along with a wide variety of injection techniques. Knowledge of the various methods and capabilities help the attacker unearth practical attacks. Otherwise, there is always a chance of developing fault attacks which are not practical, and hence not useful!
- **Classical Fault Analysis of Public-Key and Symmetric-Key Ciphers**: Different variations of fault analysis have been developed on public- and symmetric- key ciphers. The most popular form of fault analysis is what is called as Differential Fault Analysis (DFA), which is a combination of Differential Cryptanalysis and faults. The book provides a background on DFA, starting with the classic attacks on RSA. However, more detailed treatment is provided on symmetric cipher standards, like AES, when the fault affects both the datapath and key schedule. The book also provides treatment of fault analysis of stream cipher standards, like Grain, with practical results to demonstrate how real-life faults manifest and can be exploited.
- **Combination of Side-Channel and Fault Analysis**: Fault Analysis has been inspired and supplemented by side-channel analysis. Combinations have led to powerful attack vectors, like Differential Fault Intensity Attacks (DFIA), which try to utilize the fact that the fault injections are not necessarily uniform, and thus leaves a bias. DFIA present a side channel akin to analysis methodology to exploit this bias. On the other hand, researchers have tried to combine side-channel leakage, through say power consumption, and perform subsequent fault analysis to develop very strong attacks on standard cryptosystems. In the book, we support the theory with accompanied case studies on AES and PRESENT like standard ciphers.
- **Laser-based Fault Injection Techniques**: Controllability of fault injection methods is central to the success of fault attacks. Lasers provide a unique capability to target fault injections with greater accuracy, but also requires proper processing of the device and also understanding of the fault model. The book presents case studies on AES and the recently popular stream cipher, called ChaCha, to show a new type of fault attack, namely Instruction Skip Attack and Instruction Replacement Attack.
- **Software-Triggered Fault Analysis: RowHammer Bugs**: Recent day (DRAM) memory chips manifest a reliability issue, reported as RowHammers, which shows bit flips in rows adjacent to those rows which are accessed faster than the refresh rates repeatedly. This bug seems to offer a mechanism of launching fault injections through software codes! The book provides a detailed case study on a 1024-bit RSA key-based ciphering using the standard GNU-MP big integer library. It shows that though difficult, it is indeed possible to perform bit flips in the secret keys, a single instance of which is enough to reveal the complete key due to the power of DFA.

- **Automation of Fault Analysis**: Though most of the fault analysis techniques on ciphers, like AES, had been developed like the conventional cryptanalysis community, of human observation and analysis, it is much coveted to develop automated tools in this direction. An overview of such an automated method, called as Algebraic Fault Analysis (AFA) is provided with case studies on PRESENT, is provided. Such techniques indeed may pave ways to future analysis and design tools to unearth new attacks without human intervention, and thus shorten design time and effort.
- **Classical Fault Tolerance**: Several approaches for mitigating the powerful DFA have been developed using classical fault tolerance techniques. These techniques, which are largely based on various forms of redundancy, need to be understood for performing trade-offs between performance and security versus fault analysis for cryptosystems.
- **Countering Biased Fault Attacks**: The difference between classical fault tolerance and fault attacks is the fault injection methodology. In classical fault tolerance, while it is widely assumed all faults are equally likely, in attacks like DFIA there is a bias in the injector. This can lead to the increase in the probability of fault collisions, and attacks against classical fault tolerance techniques. In this context, we propose a technique called Fault Space Transformation (FST), which can be used to counter this increase of fault collision probability by changing the fault space. This technique has been illustrated with experimental results on FST being applied to AES-128.
- **Infective Countermeasures**: While the previous mitigation techniques attempt to detect a fault injection by an explicit comparison step, there is another family of countermeasures which infects the differential by a value which is not related to the key. These techniques are called as Infective Countermeasures, which have been cryptanalyzed several times and are difficult to construct. In this chapter, we present an infective countermeasure for AES-128, and provide a formal analysis method to show the level of security against various fault models. Finally, we fortify the method against the instruction fault model on an x86 ISA, using the idea of idempotent instructions. However, the security comes with the cost of a significant performance overhead which is also discussed.
- **Reactive Countermeasures**: As mentioned, all the above countermeasures come with a significant cost. In this book, we develop a new class of countermeasures, which are largely reactive in nature. The principle is based on digital artefacts and sensors being deployed in the chip, to be alert against a fault injection. A detailed case study on such a design style has been furnished on AES with reports on fault injections. This method seems to provide a very low overhead method for thwarting fault attacks by eliminating them at source.

The book thus covers a wide range of topics on fault analysis of cryptosystems, and is aimed at catering to postgraduate students and practitioners in the area of Hardware Security. It can also be used in final year graduate courses, albeit leaving out few chapters. A suggested sequence of reading the book for a beginner would be: $1 \rightarrow 2 \rightarrow 3 \rightarrow 8 \rightarrow 11$, while for a more advanced reader the book is expected to be studied in entirety. We had tried our best to reduce the mistakes in the book, however we would be grateful if you report us via emails for any pending issues.

Happy Reading!

Kharagpur, India
August 2017

Sikhar Patranabis
Debdeep Mukhopadhyay

Computer Architecture and Design Methodologies

Series editors

Twilight zone of Moore's law is affecting computer architecture design like never before. The strongest impact on computer architecture is perhaps the move from unicore to multicore architectures, represented by commodity architectures like general purpose graphics processing units (gpgpus). Besides that, deep impact of application-specific constraints from emerging embedded applications is presenting designers with new, energy-efficient architectures like heterogeneous multi-core, accelerator-rich System-on-Chip (SoC). These effects together with the security, reliability, thermal and manufacturability challenges of nanoscale technologies are forcing computing platforms to move towards innovative solutions. Finally, the emergence of technologies beyond conventional charge-based computing has led to a series of radical new architectures and design methodologies.

The aim of this book series is to capture these diverse, emerging architectural innovations as well as the corresponding design methodologies. The scope will cover the following.

Heterogeneous multi-core SoC and their design methodology

Domain-specific Architectures and their design methodology

Novel Technology constraints, such as security, fault-tolerance and their impact on architecture design

Novel technologies, such as resistive memory, and their impact on architecture design

Extremely parallel architectures

More information about this series at http://www.springer.com/series/15213

Acknowledgements

The authors would like to thank the following colleagues and fellow researchers for their graciously kind contributions, that have imparted immense value to the technical content of the book:

- Ms. Sarani Bhattacharya, Department of Computer Science and Engineering, IIT Kharagpur
- Mr. Sayandeep Saha, Department of Computer Science and Engineering, IIT Kharagpur
- Mr. Abhishek Chakraborty, Department of Computer Science and Engineering, IIT Kharagpur
- Dr. Jakub Breier, Temasek Laboratories, Nanyang Technological University, Singapore
- Dr. Shivam Bhasin, Temasek Laboratories, Nanyang Technological University, Singapore
- Dr. Dirmanto Jap, Temasek Laboratories, Nanyang Technological University, Singapore
- Dr. Wei He, Shield Lab, Central Research Institute, Huawei International Pte. Ltd., Singapore
- Dr. Chien-Ning Chen, Independent Researcher (Formerly with Division of Mathematical Sciences, School of Physical and Mathematical Sciences, NTU Singapore)

Debdeep would like to thank his parents for their constant blessings. He also expresses his gratitude to Prof. P. P. Chakrabarti for his constant guidance, encouragement, and wonderful technical ideas in spite of his busy schedule. He would also like to express his thanks to Prof. Anupam Chattopadhyay for encouraging him to write the book during his stay at NTU Singapore. He expresses his sincere thanks to all the contributory authors and collaborators, and, in particular, his student Sikhar Patranabis, for making the book see the light of the day. He would, in particular, like to mention the support of his wife for tolerating him and to his daughter Debanti for being the wind beneath his wings to fly high. Also, last but not the least, he would like to thank his research students for having faith in him and his ideas.

Sikhar would like to thank his supervisor Dr. Debdeep Mukhopadhyay for his constant encouragement, motivation, and guidance. He also thanks his colleagues at the Secured Embedded Architecture Lab, Department of CSE, IIT Kharagpur for their inputs and assistance. He expresses his gratitude to his parents and uncle for their blessings and unwavering support. He would like to thank his maternal aunt and her husband, along with his cousin Ahanaa, for their support and encouragement. Finally, he would like to mention the role of his girlfriend Ria in being a constant source of support through the many ups and downs over the years.

Contents

Part I
Fault Attacks: A Preamble

Chapter 1
Introduction to Fault Attacks

Sikhar Patranabis and Debdeep Mukhopadhyay

Abstract The advent of cloud computing and IoT have heralded an era of unprecedented levels of embedded technology and device connectivity across the globe. However, this raises concerns related to security and privacy. While one alleviate such concerns is to resort to the use of cryptographic modules for secure computation, cryptographic implementations themselves could be subjected to implementation attacks such as side-channel and fault injection analysis. In this chapter, we provide the readers with some introductory material on fault attacks, which should be helpful in understanding the subsequent chapters and technical discussions in the book.

1.1 Introduction

The progress of computing and networking technology has heralded an era where our gadgets and devices are embedded with electronics, software, sensors, actuators, and connectivity that enable these objects to collect and exchange data at all times from anywhere across the globe. In short, smarter technology spawns smarter devices, suited for smarter applications with unprecedented levels of productivity. The advent of the Internet of Things (IoT) and cloud computing has indeed revolutionized the very infrastructure of information technology, continually pushing at the boundaries of integrating the physical world into computer-based systems. IoT, for instance, allows *objects* (more specifically, a range of embedded devices) to be sensed or controlled remotely across existing network infrastructure, so that they may be integrated into full-fledged *cyber-physical systems*, encompassing technologies such as smart grids, virtual power plants, smart homes, intelligent transportation and smart cities. Cloud computing, on the other hand, provides a model for enabling ubiquitous, on-demand access to a shared pool of configurable computing resources (e.g., computer networks, servers, storage, applications and services) which can be

S. Patranabis (✉) · D. Mukhopadhyay
Secured Embedded Architecture Lab, IIT Kharagpur, Kharagpur, India
e-mail: sikhar.patranabis@iitkgp.ac.in

S. Patranabis and D. Mukhopadhyay (eds.), *Fault Tolerant Architectures for Cryptography and Hardware Security*, Computer Architecture and Design Methodologies, https://doi.org/10.1007/978-981-10-1387-4_1

rapidly provisioned and released with minimal management effort. Cloud computing and storage solutions provide users and enterprises with various capabilities to store and process their data in either privately owned, or third-party data centers that may be located far from the user–ranging in distance from across a city to across the world. In short, we are living in a world today that is equipped with empowering technology, allowing us to enjoy the benefits thereof.

However, just as a rose has its thorns, technology - even the best of it - has its pitfalls. In today's world of data intensive computation-driven services, security and data privacy is perhaps the largest of these pitfalls. According to the Business Insider Intelligence Survey conducted in the last quarter of 2014, 39% of the respondents said that security is the biggest concern in adopting Internet of things technology. In particular, as the Internet of things spreads widely, *cyber attacks are likely to become an increasingly physical (rather than simply virtual) threat*. The ubiquitous presence of embedded devices connected to the network, which is the very idea of IoT, provides an ideal opportunity for malicious adversaries to take physical control of some of them, and steal sensitive information that should ideally remain private. Although embedded systems are easy targets of attack, their security lags far behind that of, say, PCs and servers. Embedded system security is at about the same stage that PC security was in the 1990s, when the Internet was only starting to become commercialized. That said, as embedded systems have grown more complex and integral to the IoT, cloud computing and a host of other technological services, vendors are now scrambling to incorporate security into designs meant primarily for performance and efficiency.

One way to alleviate such concerns is to resort to the use of cryptographic modules for secure computation in such devices. Since a lot of embedded devices are highly resource-constrained with low processing power and memory availability, this is not always straightforward. There, however, exists a whole branch of cryptography, often described as *lightweight cryptography*, dedicated to the development of new cryptographic primitives that are highly resource-efficient, and are hence suited for embedded platforms. Since such primitives are provably secure and mathematically robust against a wide variety of attacks, they are, in theory, sufficient to secure most applications. But *theory* is so often different from *practice*, with many a sip between the cup and the lip. One such *sip* that we will focus on are implementation-based attacks on embedded devices with the ability to weaken even cryptographic primitives that are world-wide standards. More specifically, we will focus on a class of implementation-based attacks - popularly nomenclatured as *fault analysis attacks* or *fault injection attacks*.

1.2 Fault Attacks and Fault Models

We present some introductory material on fault attacks for readers not already familiar with them. The seminal work of Boneh et al. [36] demonstrated a fault attack on the popularly used RSA cryptosystem. Since then, fault attacks have been studied on a wide variety of cryptosystems, including block ciphers such as AES. So what really is a fault attack? Quite simply, it is a malicious aberration in the normal execution of the target cryptographic algorithm. This aberration often leads to an additional information leakage, which can then be used to try and recover the key. The complexity of injecting a fault and subsequently recovering the key depend on a number of factors, including: the nature of the fault, the spatio-temporal characteristics of the fault, and the fault propagation characteristics for the target algorithm. Examples of fault nature include bit flips, byte faults, stuck-at faults or random faults. In case of block ciphers, the spatio-temporal characteristics of the fault includes the precise location and round in which the fault is injected. Finally, the propagation characteristics with respect to block ciphers depend both on the cipher algorithm as well as the fault timing. In general, based on the attack principle, fault attacks on block ciphers may be categorized into the following major categories:

1.2.1 Differential Fault Analysis (DFA)

DFA [99, 131, 146, 154, 172] is a fault analysis technique in which the adversary injects a fault with a fixed spatio-temporal characteristic, and then analyzes the differential of fault-free and faulty ciphertext pairs to recover the secret-key. DFA of several popular block ciphers have been widely reported - most widely for AES-128. In fact, state-of-the-art DFA techniques allow recovery of the entire 128 bit key of AES with as few as a *single* fault injection [172]. DFA usually uses practically achievable fault models that can be injected using low cost fault injection techniques [172].

1.2.2 Fault Sensitivity Analysis

Fault sensitivity analysis, abbreviated as FSA, was introduced by Li et al. in [110]. FSA exploits the data-dependent transition point between the fault-free and faulty behavior of a given target cipher implementation. For example, adversaries may discern the critical point of fault occurrence while gradually increasing the intensity of the fault injection. FSA then exploits the relationship between the fault sensitivity and the processed sensitive data to retrieve the secret information from a cryptographic device.

1.2.3 Differential Fault Intensity Analysis (DFIA)

DFIA [67] represents a class of fault attacks that combine principles of side channel analysis techniques such as DPA with that of fault analysis for key-recovery [66].

DFIA requires *only* faulty ciphertexts, and chooses the key hypothesis that can be traced back to a faulty intermediate state value with a highly biased distribution, as opposed to a uniformly random distribution. DFIA requires higher number of fault models owing to its inherently statistical nature; however, it is a potent threat to the security of several block ciphers.

1.2.4 Safe-Error Attacks (SEA) and Differential Behavior Analysis (DBA)

The final class of fault attacks reported in the cryptographic literature is the safe error attack (SEA) and differential behavior analysis (DBA) that deduce from the *behavior* of a cryptographic device whether a fault leading to an incorrect computation has actually occurred during an encryption [32, 155]. The attack exploits the data-dependent fault propagation characteristics of a block cipher under an appropriate choice of fault model. In particular, whether or not a fault injection would result in a faulty ciphertext often depends on one or more segments of the secret-key; a typical example being the attack on an AES implementation targeting the **xtime** operation [32]. This attack reveal secret key bytes based on the response of the device to fault injections. More recent propositions in this direction include the fault behavior analysis (FBA) [109]. These attack only observe the behavior of the device under fault injections, and do not require the value of the ciphertext. However, they tend to use slightly stringent fault models such as stuck-at faults, which are costly to achieve in practice.

1.3 Fault Injection Techniques: Semi-invasive and Non-invasive Methodologies

Over the years, security experts have demonstrated a number of possible mechanisms via which faults may be injected during the cryptographic operation of smartcard ICs and FPGAs. Here, we introduce to the readers a few non-invasive and semi-invasive fault injection techniques that have been popularly studied over the years. Typically, a convenient fault injection technique should avoid costly operations such as physically opening a device, application of chemical preparations to improve leakages, or electrical contact to the device surface. We present four such convenient fault injection techniques below:

1.3.1 Spike Attacks

A spike is a sudden deviation of the supply voltage/power levels to a smartcard IC/FPGA beyond its pre-specified tolerance limits. For example, 5 V is the standard maximum voltage that can be supplied to a smart card for it to function properly. A sudden spike in the voltage supply level beyond 5 V would therefore lead to faulty computations, with the nature of the fault determined by the combined effect of the spike timing, spike voltage value and the spike voltage transition shape. These parameters can be appropriately adjusted by an adversary to obtain a spike that injects the desired fault in the working of a cryptographic algorithm.

1.3.2 Glitch Attacks

Similar to spike attacks, glitch attacks typically use a sudden deviation of the supply clock to a circuit implemented on a smartcard IC/FPGA. Once the glitch is beyond the tolerable clock frequency (in other words, the minimum tolerable clock rise and clock fall time) for the circuit to produce functionally correct output, it leads to fault injections. Clock glitches are most commonly used to create setup time violations in hardware implementations on FPGAs, and to skip/replace instructions during execution of programs on software platforms. Since the advent of fault attacks, clock glitches have been the cheapest and most popularly used form of fault injection technique on cryptographic ICs and hardware [8, 9, 32, 105].

1.3.3 Optical/Laser Attacks

Optical attacks were originally used to inject faults in the non-volatile memory or the EEPROM of smartcard ICs [8–10, 30, 104, 105, 115, 134, 144]. Subsequent advances, however, made it possible to inject more targeted faults in specific locations of a micro-controller using highly focused camera flash-light/laser beams. State-of-the-art attack techniques have the ability to set/reset one or more memory/register bits in the microcontroller. A direct application of this fault injection technique is to affect the opcode bits of the instruction register in a microcontroller, resulting in instruction skips/replacements with high precision in timing and high repeatability. Equipment for such attacks range from cheap publicly available photographic equipment to advanced laser stations. One of the disadvantage of such attacks are that they are semi-invasive in nature and sometimes require physical openings in the IC to work.

1.3.4 Electromagnetic Attacks

Electromagnetic (EM) attacks [150] can be used to set/reset individual bits in micro-controller memory/FPGA registers by inducing eddy currents in the target locations. The usual technique is to place a coil conducting alternating current near the target location on a microcontroller surface. The attack results in transient faults at the bit-level with high spatio-temporal controllability, and again requires cheap publicly available equipment. For example, one can simply use a signal generator to generate the alternate current of appropriate magnitude, which may then be connected electrically to a coil of appropriate dimensions for creation of the EM field that induces eddy currents in the target device. In some cases, the intensity of the EM pulse may be enhanced via an amplifier. It is a fully non-invasive technique that works easily through the plastic covering of a smartcard [3, 61, 125, 149].

1.4 Organization of the Book

The rest of the book is organized as follows. Chapter 2 presents classical differential fault analysis-based fault attacks on public and private-key cryptosystems. Chapters 3 and 4 present slightly unconventional fault attacks on block ciphers that are either inspired by or assisted by side-channel analysis techniques. Chapter 5 introduces to the readers a very specialized fault injection technique targeting microcontrollers using laser pulses, along with case studies to illustrate its effectiveness. Chapter 7 provides the readers with a novel perspective on automation of fault analysis from a design point of view. The rest of the chapters deal with different countermeasure techniques to prevent fault attacks. Chapter 8 presents classical concurrent error detection techniques for detection and prevention of traditional DFA attacks. Chapter 9 presents fault space transformation - a recent countermeasure technique that works well even when the underlying fault model is non-uniform and controllable by the adversary. Chapter 10 presents a third class of countermeasures, known as infective countermeasures, that is provably secure against first order DFA. This chapter also uses the infective countermeasure as a case study to introduce a second class of countermeasures, called instruction-level redundancy, that is particularly designed to prevent skip attacks on software platforms such as microcontrollers. Finally, Chap. 11 presents sensor-based reactive strategies to prevent specialized attack techniques such as laser pulse injections.

Chapter 2
Classical Fault Attacks on Public and Symmetric-Key Cryptosystems

Sikhar Patranabis, Abhishek Chakraborty and Debdeep Mukhopadhyay

Abstract This chapter presents to the readers some classical fault attacks on both public as well as symmetric-key cryptosystems. We begin with the celebrated fault analysis of the decryption algorithm in RSA. This is followed by an ensemble of fault analysis techniques targeting the Advanced Encryption Standard (AES). We end with a more recent fault attack targeting Grain-128 - an eSTREAM finalist and popularly used stream cipher. Nearly all the fault attacks described in this chapter can be achieved with low -cost fault injection techniques such as clock/voltage glitches.

2.1 The Seminal Fault Attack on RSA

We motivate the study of fault attacks using the first such attack to be proposed in the literature - on the RSA cryptosystem [36]. It is well-known that RSA is a public-key cryptosystem operating using a public-private key pair (PK, SK) over a finite field $\mathbb{Z}_n$, containing all integers modulo n, where n is the product of a pair of large primes p and q. While n is publicly known as part of PK, p and q are secret, and are assumed to be difficult to compute from n under the assumption that factorization of product of large primes is intractable. Besides n, the public key PK includes the encryption exponent $e \in [1, \phi(n)]$, where $\phi(n) = (p-1)(q-1)$ is the Euler-Totient function. The private key SK consists of the corresponding decryption exponent $d \in [1, \phi(n)]$ such that $d = e^{-1} \bmod \phi(n)$, along with the primes p and q. The algorithm operates as follows:

- **Encryption**: The ciphertext C corresponding to a message $M \in \mathbb{Z}_n$ is computed as $C = M^e \bmod n$.
- **Decryption**: The message M is retrieved from the ciphertext C as $M = C^d \bmod n$

The security of RSA is based on the intractability of computing the decryption exponent d from the knowledge of n and the encryption exponent e, provided that

S. Patranabis (✉) · A. Chakraborty · D. Mukhopadhyay
Secured Embedded Architecture Lab, IIT Kharagpur, Kharagpur, India
e-mail: sikhar.patranabis@iitkgp.ac.in

S. Patranabis and D. Mukhopadhyay (eds.), *Fault Tolerant Architectures for Cryptography and Hardware Security*, Computer Architecture and Design Methodologies, https://doi.org/10.1007/978-981-10-1387-4_2

n is sufficiently large. This is, in turn, based on the intractability of computing $\phi(n)$ for large n, which in turn is based on the intractability of factorizing n to obtain the secret primes p and q.

In their seminal work of 1997, Boneh, DeMillo and Lipton demonstrated that any implementation of the RSA cryptosystem must check for the occurrence of faults that could otherwise leak the secret decryption exponent d. They assumed a single bit flip model, where the adversary is able to flip a specific bit of the decryption exponent d during the decryption operation on a ciphertext C. Suppose $d = d_{m-1}d_{m-2}\cdots d_0$ in binary representation, and a bit d_i is flipped to d_i' (from either 1 to 0 or from 0 to 1) by the adversary. Let M' be the corresponding faulty plaintext message recovered. It is assumed that the adversary has access to both the correct plaintext message M and the faulty message M' corresponding to the ciphertext C. The adversary can now compute $M/M' = C^{2^i \cdot (d_i - d_i')}$, and guess the value of d_i as either 1 or 0 depending on whether $M/M' = C^{-2^i}$ or $M/M' = C^{2^i}$, respectively. The same technique can be repeated over all the bit positions from 0 to $m-1$, which allows the entire secret exponent d to be retrieved. The required number of fault injections can also be reduced by flipping two or more bits at once, and using them to recover multiple bits of d in parallel.

2.2 Differential Fault Analysis of Block Ciphers

Since the seminal work of Boneh et al. targeting the RSA algorithm, a large body of fault attacks have followed a common strategy referred to as differential fault analysis (DFA in short). DFA was pioneered by Biham and Shamir in [30], who targeted the then symmetric-key encryption standard DES - a Fiestel cipher. DFA has since been applied to a wide variety of block ciphers since then, including both Fiestel-based ciphers as well as substitution-permutation network (SPN)-based ciphers such as the Advanced Encryption Standard (AES). In this section, we present to the readers the basic principles of mounting a DFA attack on block ciphers. Subsequently, we illustrate the capabilities of DFA using multiple case-studies on AES - a mathematically robust cryptosystem and the current worldwide standard for block ciphers. We will illustrate to the readers that DFA is powerful enough to recover the entire secret key of AES using just a single fault injection.

2.2.1 The Basic Principle of DFA

As the name suggests, DFA combines principles of differential cryptanalysis with fault injection to recover the secret key of a block cipher. The *differential* is usually computed between two executions of the same cryptographic algorithm on the same plaintext and using the same secret key:

1. One of these executions is fault-free and yields the correct ciphertext.
2. The other execution is perturbed by the adversary via one or more fault injections to obtain a faulty ciphertext.

It is generally assumed that the adversary has precise information about the fault location, and some general idea about the nature of the fault or the *fault model*. The attack then proceeds by analyzing the relationship between the fault-free and faulty ciphertexts using the knowledge of the fault model and fault location to recover the secret key.

2.2.2 Permanent and Transient Faults

Depending on the manner in which they occur in a target device, faults may be classified into two categories - permanent and transient. Permanent faults usually manifest when a target device is subjected to extreme perturbations over long periods of time, such as high voltage, excessive temperatures or even highly focused laser/EM pulses injected for long periods of time. Permanent faults often result in specific bits in one or more device registers losing their ability to switch, resulting in *stuck-at* conditions. From a cryptographic design perspective, this could lead to certain bit(s) in the internal states/final outputs being permanently assigned either 0 or 1, leading to a loss of entropy which could be exploited by an adversary. However, permanent faults are by nature easy to detect via simulation-based fault detection mechanisms, and are fixed by repairing/replacing the target device.

Transient faults usually offer a greater threat to cryptographic implementations since they are caused by sudden non-persistent perturbations to the target device via glitches in voltage/clock frequency or short directed pulses from an EM/ laser gun. These faults usually do not damage the device, implying that there is no after-effect of the fault beyond the target execution of the cryptographic algorithm. These faults thus achieve the same goal as permanent faults, while being more evasive against simulation-based testing and hence, more difficult to prevent.

2.2.3 Fault Models for DFA

A number of different fault models have been used in the cryptographic literature when mounting DFA attacks on cryptographic implementations. Many attacks assume random single bit faults, which are extremely targeted in the sense that they flip exactly one bit of the cipher state, and are hence usually difficult to achieve without precise fault injection equipment. They are, however, extremely useful from an adversary's perspective since they potentially lead to the maximum reduction in key entropy during analysis. Single byte faults are a more general class of faults that are also often exploited in several attack methodologies. These faults offer a good trade-off between practicality of injection and analysis efficiency, and are hence the most practically exploitable. Multiple byte fault models are the easiest to inject, and have also been exploited in certain attack instances, albeit under the condition that the affected bytes share certain common properties (for example, they should lie along the same diagonal of the cipher state matrix and so on). While these fault models could potentially make the analysis more complex, their abundant presence upon fault injection via even low-cost fault injection techniques makes them a potent threat to cryptographic implementations.

2.2.4 DFA on a Generic SPN Block Cipher

In this section, we illustrate the basic principles of DFA on a generic SPN-based block cipher illustrated in Fig. 2.1. The block cipher comprises of r rounds and a block length of n bytes. Each round consists of a non-linear substitution-box (S-Box) layer S, a linear diffusion layer D, and a combination with the round key. An additional combination with the secret key, referred to as *key-whitening* is performed at the beginning of the encryption. The diffusion layer usually consists of a permutation of the bits/bytes/nibbles of the cipher state matrix, optionally followed by a linear transformation using a maximum-distance separable (MDS) matrix. For the sake of simplicity, we assume that both the S-Box layer S and the diffusion layer D operate on the bytes of the state matrix (meaning that no operations occur within a given byte). In this setting, if byte is modified/perturbed at the input of D in a given round, the induced differential spreads to multiple bytes at the output of D in the same round. The exact differential propagation depends on the differential branch number of D, which is essentially the minimum possible sum of the number of perturbed bytes at the input and output of D. In the literature of differential cryptanalysis, perturbed bytes are also referred to as *active bytes*, while in the fault attack literature, it is more common to refer to them as *faulty bytes*. Thus the differential branch number of D puts a lower bound on the minimum number of faulty bytes at the output of D corresponding to a given number of faulty bytes at its input.

Now, say a single byte fault is induced at the input of $(r-1)$th (penultimate) round and the corresponding difference at the input of D^{r-1} is $\alpha \neq 0$. Let the branch number of the diffusion layer be b. Then, as per the discussion above, the input byte fault will spread to $b-1$ bytes $(\alpha_{\pi_0}, \ldots, \alpha_{\pi_{b-2}})$ at the output of D_{r-1}, where π denotes the transformation of the diffusion layer. Each of these faulty bytes are non-linearly transformed by the S-Boxes, and the adversary can represent the output bytes in terms of a pair of fault-free and faulty ciphertext bytes (C, C^*) as:

$$\alpha_{\pi_j} = S^{-1}(C_{\pi_j} \oplus K^r_{\pi_j}) \oplus S^{-1}(C^*_{\pi_j} \oplus K^r_{\pi_j}) \tag{2.1}$$

where $j \in \{0, \ldots, b-2\}$ and S^{-1} represent the inverse of the S-Box operation. Now, since the adversary knows the S-Box input difference, $C_{\pi_j} \oplus C^*_{\pi_j}$, he can infer from

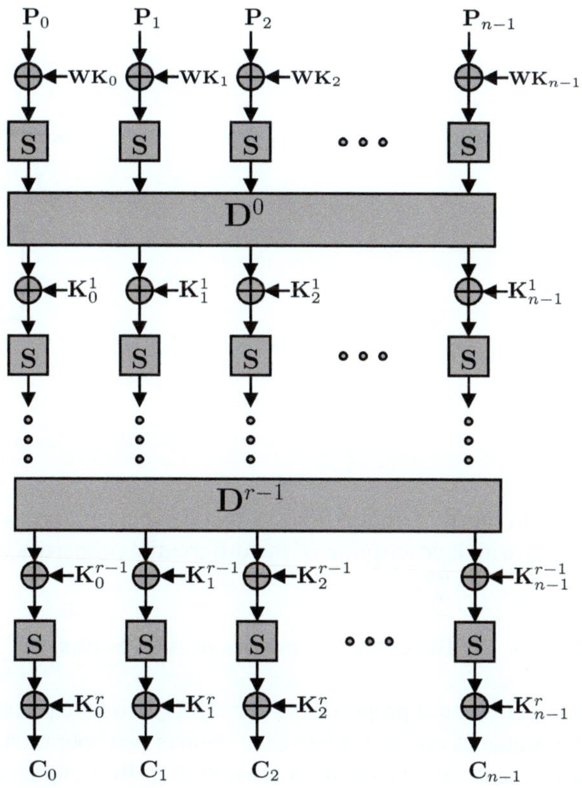

Fig. 2.1 Basic structure of SPN ciphers

the S-Box difference distribution table, the few possible values of $K^r_{\pi_j}$ satisfying the above equation. This leads to a reduction in the entropy of $K^r_{\pi_j}$, which can be further reduced via multiple attacks/brute force trials as suitable. Finally, just as in any divide-and-conquer approach, the adversary may repeat the same attack procedure to recover multiple bytes of the last round key. Since in almost any block cipher, each round key can be deterministically traced back to the original round key via a deterministic key scheduling, the attack reveals the original secret key of the cipher.

2.2.5 Case Study: Differential Fault Analysis of the Advanced Encryption Standard

In this section, we present to the readers a case study of DFA on AES using a variety of fault locations and fault models. We present several fault analysis techniques that require different number of fault injections to retrieve the secret key of AES. The

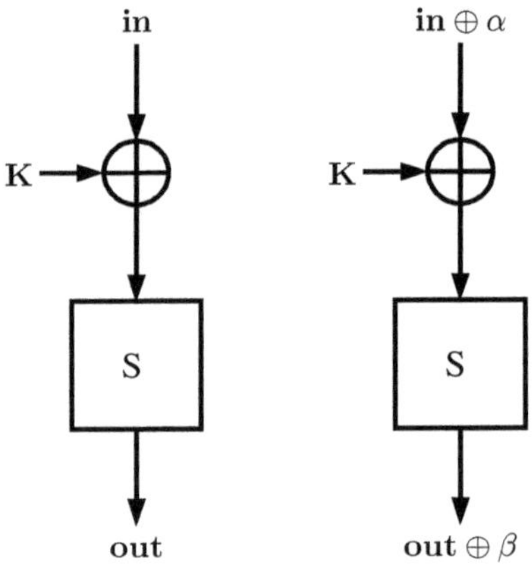

Fig. 2.2 Differential properties of the AES S-Box

attacks mostly target AES-128 that has both plaintext and key size of 128 bits. We begin with a description of the differential properties of the AES S-Box.

2.2.5.1 Differential Properties of AES S-Box

The differential properties of the AES S-Box are pivotal to the success of mounting DFA attacks on AES. Figure 2.2 depicts two operations of the AES S-Box: one with an input in and the other an input $\mathsf{in} \oplus \alpha$. Both inputs are XOR-ed with a round key byte K. The corresponding differential β is obviously non-linearly associated with the input differential α, which may be captured by the following equation:

$$S(\mathsf{in} \oplus K) \oplus S(\mathsf{in} \oplus K \oplus \alpha) = \beta \tag{2.2}$$

Abbreviating $X = \mathsf{in} \oplus K$, the equation takes the form:

$$S(X) \oplus S(X \oplus \alpha) = \beta \tag{2.3}$$

Now, as per the properties of the AES S-Box documented in the literature, for a given pair of α and β, the above equation can have either 0, 2 or 4 solutions for X [137]. In particular, for a given α, 126 out of the possible 256 values of β lead to 2 solutions for X, one value of β yields 4 solutions for X, and the remaining values yield no solution for X. Thus, over all 256 possible values of β, the average number of solutions per equation is 1. Now, from th knowledge of in and the solutions for X, one can compute the possible values of K satisfying Equation 2.2. Also, for a given α, over all 256 possible values of β, the average number of solutions for K per equation is 1. This property of the AES S-Box plays a crucial role in all the DFA attacks discussed subsequently.

Fig. 2.3 Bit-fault: last round of AES-128

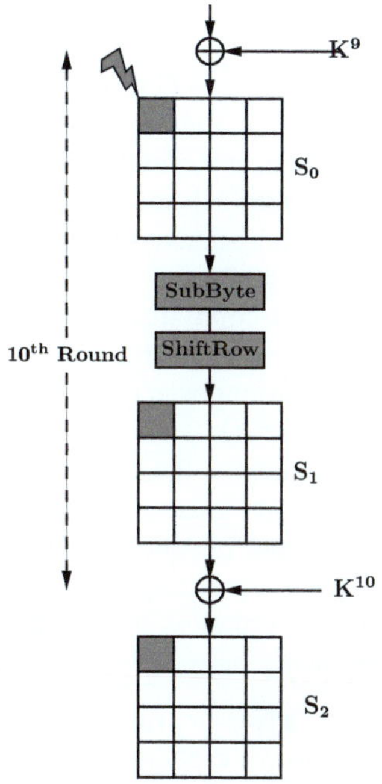

2.2.5.2 DFA of AES-128 Using Bit Faults

We now introduce to the readers two of the simplest DFA attacks on AES using single-bit faults. The attacks differ in the fault locations - the first attack targets the last round of AES, while the second targets the initial key-whitening step. Both attacks assume that the adversary has access to faulty and fault-free ciphertexts corresponding to each fault injection, and neither attack assumes that the adversary knows the exact nature of the flip induced by the fault. We now describe the two bit-flip based DFA attacks on AES-128.

The first fault attack is depicted pictorially in Fig. 2.3. The 9 and 10th round keys for AES-128 are denoted as K^9 and K^{10}, respectively. Additionally, the XOR of the fault-free and faulty intermediate states across different rounds is depicted using the state matrices S_0, S_1, and S_2. Note that the adversary injects a single bit upset; consequently, in the absence of a MixColumn operation in the the 10th round, the XOR difference of fault-free and faulty ciphertexts clearly reveal the byte where the fault was injected. Of course, the actual flipped bit is not immediately evident due to the presence of the S-Box; however, we discuss how this may be inferred. Suppose that the fault induced at the (i, j)th byte of the state matrix S_0 at the input of the 10th round. Also, let x be the byte input to the 10th round for a fault-free execution, and let ε be the aberration in the byte value after fault injection. From the XOR difference of the correct ciphertext C and the faulty ciphertext C^*, the adversary not only gets the information about the faulty byte, but also formulates the following equation:

$$C_{i,l} \oplus C^*_{i,l} = SR(S(x_{i,j})) \oplus SR(S(x_{i,j} \oplus \varepsilon)) \tag{2.4}$$

where $l = (j - i) \bmod 4$ is an adjustment made to identify the corresponding column index where the faulty byte in the jth column and ith row shifts due to the ShiftRow operation, at the tenth round output. The aforementioned technique may be replicated to recover each individual byte of the last round key. The overall number of fault injection instances required across the entire key-recovery process is expected to be $2 \times 16 = 32$.

The second attack is conceptually similar but more difficult to mount in practice. It targets the initial key-whitening operation of AES-128 using a stuck-at fault model, as opposed to the bit flip model assumed in the previous attack. The plaintext for the attack is chosen to be all zeros, and is kept constant throughout the attack. Consequently, the key-whitening step ensures that the input to the first round is the same as the master secret-key. It is in this first round input state that the adversary injects multiple bit-wise stuck-at faults. The assumption that an adversary is capable of precisely setting/resetting a given bit of an internal cipher state is rather restrictive; nevertheless the resulting attack is simple and instructive, as described in the following steps:

- The adversary runs a fault-free encryption on the all-zero plaintext and obtains the fault-free ciphertext C.
- The adversary now re-runs the encryption on the same plaintext, and performs a stuck-at-zero fault injection on a specific bit of the cipher state after the initial key-whitening and prior to the first round SubByte operation. Let C' be the corresponding faulty ciphertext.
- If $C = C'$, it implies that the fault injection had no effect on the overall encryption algorithm; hence the target bit of the master secret-key must have been originally 0. on the other hand, if $C \neq C'$, this bit must have been 1 originally.

The above procedure must be repeated individually for each of the 128 possible bit locations. Achieving these many bit-reset faults within a practically feasible number of fault injections is the most difficult part of the aforementioned attack. Nonetheless, the attack is a good illustration of the power of fault analysis in general, at least in theory.

2.2.5.3 DFA of AES-128 Using Byte Faults

As already discussed, byte faults are a superclass of bit faults, and are hence more practically achievable. In a byte fault, any number of bits within a single byte can get flipped; however, the effect of the fault must not spread beyond a single byte. We now present to the readers a byte-fault based DFA of AES that targets the penultimate round. In this attack, the adversary injects a byte fault between the MixColumn operations in the 8 and 9th round of AES encryption, and exploits the differential properties of the AES S-Boxes, along with the diffusion properties of the MixColumn operation, to recover multiple key bytes in parallel. Figure 2.4 illustrates the propagation of the single byte fault induced in the penultimate round to four different bytes in the final round, owing to the differential branch number of the MixColumn transformation. The induced fault has generated a single byte difference at the input of the 9th round MixColumn. Let f be the byte value of the difference and the corresponding 4-byte output difference is $(2f, f, f, 3f)$, where 2, 1, and 3 are the elements of the first row of the MixColumn matrix. The 4-byte difference is again converted to (f_0, f_1, f_2, f_3) by the non-linear S-Box operation in tenth round. The ShiftRow operation will shift the differences to 4 different locations. The adversary has access to the fault-free ciphertext C and faulty ciphertext C^*, which differs only in 4 bytes. Now, we can represent the 4-byte difference $(2f, f, f, 3f)$ in terms of the tenth round key K^{10} and the fault-free and faulty ciphertexts by the following equations:

$$\begin{aligned} 2f &= S^{-1}(C_{0,0} \oplus K^{10}_{0,0}) \oplus S^{-1}(C^*_{0,0} \oplus K^{10}_{0,0}) \\ f &= S^{-1}(C_{1,3} \oplus K^{10}_{1,3}) \oplus S^{-1}(C^*_{1,3} \oplus K^{10}_{1,3}) \\ f &= S^{-1}(C_{2,2} \oplus K^{10}_{2,2}) \oplus S^{-1}(C^*_{2,2} \oplus K^{10}_{2,2}) \\ 3f &= S^{-1}(C_{3,1} \oplus K^{10}_{3,1}) \oplus S^{-1}(C^*_{3,1} \oplus K^{10}_{3,1}) \end{aligned} \tag{2.5}$$

Each of the above equations can now individually be represented in the form of Eq. 2.2, with an average of one solution per value of f. Note that taken in isolation, this does not reduce the entropy of the key bytes involves as f can now take all 2^8 values. However, observe that we now have 5 unknown bytes - namely the fault value f and the four key bytes - $K^{10}_{0,0}$, $K^{10}_{1,3}$, $K^{10}_{2,2}$ and $K^{10}_{3,1}$, across 4 equations, instead of 2 unknown values in a single equation. Consequently, the joint search space of all the variables is reduced to $\left(2^8\right)^{5-4}$, which is only 2^8. Thus, in a single fault injection, the joint search space for the key bytes $K^{10}_{0,0}$, $K^{10}_{1,3}$, $K^{10}_{2,2}$ and $K^{10}_{3,1}$ is reduced from 2^{32} to 2^8. On an average, two fault injections should therefore be sufficient to reduce the search space to a unique quartet of values. The same procedure may be repeated by injecting single byte faults in the 3 other columns of the state matrix to recover the remaining 12 key bytes - four at a time per fault injection. Thus the overall attack requires 8 fault injections to recover all 16 bytes of the final round key.

The attack described upon can be further improved upon to reduce the number of fault injections necessary for recovering the secret key. Suppose that the same byte fault was injected in the AES state matrix one round earlier, that is, between the MixColumn operations in the 7 and 8th round of AES encryption. Figure 2.5 depicts the propagation of faults in this case. The MixColumn operation in round 8 ensures that the fault propagates to four different bytes in the same column of the state matrix, while the ShiftRow operation in round 9 propagates each fault to a different column of the state matrix. We now see this is exactly equivalent to having four parallel instances of the single byte fault described above. Each of these four resulting faults can now be analyzed individually to reduce the search space of different quartets of the last round key. On an average, therefore, the number of fault injections required is reduced from 8 to 2. An attack following this principle was reported in [71], which

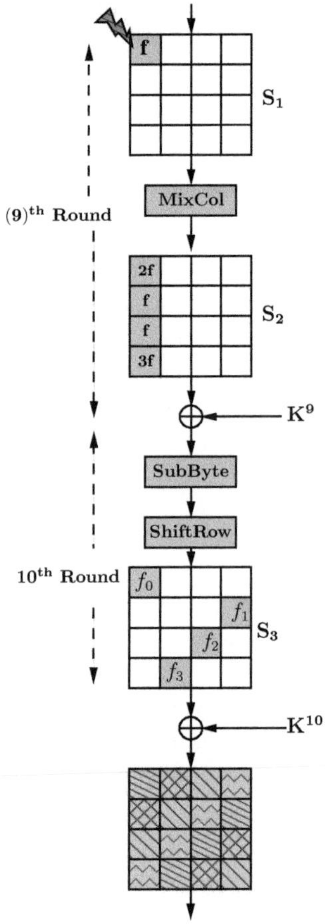

Fig. 2.4 Fault propagation across the last two rounds of AES-128

is the best known fault attack till data on AES-128 under the assumption that the adversary does not have control over the plaintext being encrypted.

2.2.5.4 DFA of AES-128 Using a Single Byte Fault

We now present to the readers a DFA of AES-128 that uses a single byte fault injection to recover the entire 128-bit secret key. In this attack, the adversary is assumed to have control over the plaintext being encrypted. Besides this, the attack uses the same fault model and location as the previous attack. The main advantage of this attack over the previous one is that upon a single fault injection, the adversary can reduce the key space to the extent that a second fault injection is no longer necessary and a brute force attack can uniquely recover the key. Note that in a practical fault injection scenario, both attacks will require more fault injections than the theoretical

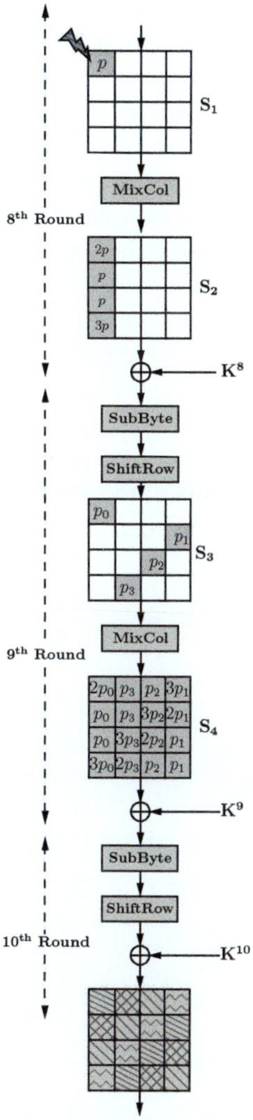

Fig. 2.5 Fault propagation across the last three rounds of AES-128

estimates; hence a more pragmatic view of this attack is that it reduces the required number of fault injections to around half of that in the previous attack.

We revisit the fault propagation characteristics depicted in Fig. 2.5. From the first column of S_4 we get following 4 differential equations:

$$\begin{aligned} 2\,p_0 &= S^{-1}(C_{0,0} \oplus K^{10}_{0,0}) \oplus S^{-1}(C^*_{0,0} \oplus K^{10}_{0,0}) \\ p_0 &= S^{-1}(C_{1,3} \oplus K^{10}_{1,3}) \oplus S^{-1}(C^*_{1,3} \oplus K^{10}_{1,3}) \\ p_0 &= S^{-1}(C_{2,2} \oplus K^{10}_{2,2}) \oplus S^{-1}(C^*_{2,2} \oplus K^{10}_{2,2}) \\ 3\,p_0 &= S^{-1}(C_{3,1} \oplus K^{10}_{3,1}) \oplus S^{-1}(C^*_{3,1} \oplus K^{10}_{3,1}) \end{aligned} \quad (2.6)$$

Since p_0 can take 2^8 possible values in the aforementioned system of equations, the overall search space for the entire quartet of key bytes is also reduced to 2^8 (this follows from the S-Box difference distribution table for AES-128). Repeating this procedure independently for all four columns results in a set of 2^8 possible values for each key-quartet, and a combined search space of 2^{32} for the entire last round key K^{10}. Further, if the fault location is not known apriori to the adversary, the search space for the entire last round key K^{10} would be at most $2^{32} \times 16 = 2^{36}$. This search space can be enumerated and tested exhaustively within practically feasible time limits.

Note that the aforementioned key-recovery process left unexplored the relation between the fault values at the state matrix S_2. Bringing these relations into the fold facilitates reducing the key search space even further. Suppose that the adversary knows the fault location apriori, and obtains a total of 2^{32} possible key hypotheses from the first phase of the attack as described above. The fault value in the first column of the state matrix S_2 can be represented in terms of K^9 (the 9th round key), C^9 (the 9th round fault-free output) and C^{*9} (the 9th round faulty output) as follows:

$$
\begin{aligned}
2\,p_0 = S^{-1}(&14(C^9_{0,0} \oplus K^9_{0,0}) \oplus 11(C^9_{1,0} \oplus K^9_{1,0}) \oplus \\
&13(C^9_{2,0} \oplus K^9_{2,0}) \oplus 9(C^9_{3,0} \oplus K^9_{3,0})) \oplus \\
S^{-1}(&14(C^{*9}_{0,0} \oplus K^9_{0,0}) \oplus 11(C^{*9}_{1,0} \oplus K^9_{1,0}) \oplus \\
&13(C^{*9}_{2,0} \oplus K^9_{2,0}) \oplus 9(C^{*9}_{3,0} \oplus K^9_{3,0}))
\end{aligned} \tag{2.7a}
$$

$$
\begin{aligned}
p_0 = S^{-1}(&9(C^9_{0,3} \oplus K^9_{0,3}) \oplus 14(C^9_{1,3} \oplus K^9_{1,3}) \oplus \\
&11(C^9_{2,3} \oplus K^9_{2,3}) \oplus 13(C^9_{3,3} \oplus K^9_{3,3})) \oplus \\
S^{-1}(&9(C^{*9}_{0,3} \oplus K^9_{0,3}) \oplus 14(C^9_{1,3} \oplus K^{*9}_{1,3}) \oplus \\
&11(C^{*9}_{2,3} \oplus K^9_{2,3}) \oplus 13(C^{*9}_{3,3} \oplus K^9_{3,3}))
\end{aligned} \tag{2.7b}
$$

$$
\begin{aligned}
p_0 = S^{-1}(&13(C^9_{0,2} \oplus K^9_{0,2}) \oplus 9(C^9_{1,2} \oplus K^9_{1,2}) \oplus \\
&14(C^9_{2,2} \oplus K^9_{2,2}) \oplus 11(C^9_{3,2} \oplus K^9_{3,2})) \oplus \\
S^{-1}(&13(C^{*9}_{0,2} \oplus K^9_{0,2}) \oplus 9(C^{*9}_{1,2} \oplus K^9_{1,2}) \oplus \\
&14(C^{*9}_{2,2} \oplus K^9_{2,2}) \oplus 11(C^{*9}_{3,2} \oplus K^9_{3,2}))
\end{aligned} \tag{2.7c}
$$

$$
\begin{aligned}
3\,p_0 = S^{-1}(&13(C^9_{0,1} \oplus K^9_{0,1}) \oplus 9(C^9_{1,1} \oplus K^9_{1,1}) \oplus \\
&14(C^9_{2,1} \oplus K^9_{2,1}) \oplus 11(C^9_{3,1} \oplus K^9_{3,1})) \oplus \\
S^{-1}(&13(C^{*9}_{0,1} \oplus K^9_{0,1}) \oplus 9(C^{*9}_{1,1} \oplus K^9_{1,1}) \oplus \\
&14(C^{*9}_{2,1} \oplus K^9_{2,1}) \oplus 11(C^{*9}_{3,1} \oplus K^9_{3,1}))
\end{aligned} \tag{2.7d}
$$

In order to utilize the above equations we need the 9th-round key. The 9th-round key can be derived from the final round key by the following conversion matrix:

$$\begin{pmatrix} (K^{10}_{0,0} \oplus S[K^{10}_{1,3} \oplus K^{10}_{1,2}] & K^{10}_{0,1} \oplus K^{10}_{0,0} & K^{10}_{0,2} \oplus K^{10}_{0,1} & K^{10}_{0,3} \oplus K^{10}_{0,2} \\ \oplus h_{10}) & & & \\ (K^{10}_{1,0} \oplus S[K^{10}_{2,3} \oplus K^{10}_{2,2}]) & K^{10}_{1,1} \oplus K^{10}_{1,0} & K^{10}_{1,2} \oplus K^{10}_{1,1} & K^{10}_{1,3} \oplus K^{10}_{1,2} \\ (K^{10}_{2,0} \oplus S[K^{10}_{3,3} \oplus K^{10}_{3,2}]) & K^{10}_{2,1} \oplus K^{10}_{2,0} & K^{10}_{2,2} \oplus K^{10}_{2,1} & K^{10}_{2,3} \oplus K^{10}_{2,2} \\ (K^{10}_{3,0} \oplus S[K^{10}_{0,3} \oplus K^{10}_{0,2}]) & K^{10}_{3,1} \oplus K^{10}_{3,0} & K^{10}_{3,2} \oplus K^{10}_{3,1} & K^{10}_{3,3} \oplus K^{10}_{3,2} \end{pmatrix}.$$

Thus, for each hypothesis of K^{10} not ruled out in the first phase of the attack, and for the fault-free and faulty ciphertext pair (C, C^*), the adversary can obtain a unique triplet of the form (K^9, C^9, C^{*9}). These can then be tested using the aforementioned system of equations. The adversary then tests the aforementioned system of equations. If a given hypothesis of K^{10} fails to yield a satisfying triplet, it is immediately rejected. For completeness, we state the detailed equations as follows:

$$\begin{aligned} 2p_0 = S^{-1}\big[&14(S^{-1}[K^{10}_{0,0} \oplus C_{0,0}] \oplus K^{10}_{0,0} \oplus S[K^{10}_{1,3} \oplus K^{10}_{1,2}] \oplus h_{10}) \oplus \\ &11(S^{-1}[K^{10}_{1,3} \oplus C_{1,3}] \oplus K^{10}_{1,0} \oplus S[K^{10}_{2,3} \oplus K^{10}_{2,2}]) \oplus \\ &13(S^{-1}[K^{10}_{2,2} \oplus C_{2,2}] \oplus K^{10}_{2,0} \oplus S[K^{10}_{3,3} \oplus K^{10}_{3,2}]) \oplus \\ &9(S^{-1}[K^{10}_{3,1} \oplus C_{3,1}] \oplus K^{10}_{3,0} \oplus S[K^{10}_{0,3} \oplus K^{10}_{0,2}])\big] \oplus \\ S^{-1}\big[&14(S^{-1}[K^{10}_{0,0} \oplus C^*_{0,0}] \oplus K^{10}_{0,0} \oplus S[K^{10}_{1,3} \oplus K^{10}_{1,2}]) \oplus \\ &11(S^{-1}[K^{10}_{1,3} \oplus C^*_{1,3}] \oplus K^{10}_{1,0} \oplus S[K^{10}_{2,3} \oplus K^{10}_{2,2}]) \oplus \\ &13(S^{-1}[K^{10}_{2,2} \oplus C^*_{2,2}] \oplus K^{10}_{2,0} \oplus S[K^{10}_{3,3} \oplus K^{10}_{3,2}]) \oplus \\ &9(S^{-1}[K^{10}_{3,1} \oplus C^*_{3,1}] \oplus K^{10}_{3,0} \oplus S[K^{10}_{0,3} \oplus K^{10}_{0,2}])\big]. \end{aligned} \tag{2.8}$$

Similarly, the other 3 faulty bytes can be expressed by the following equations:

$$\begin{aligned} p_0 = S^{-1}\big[&14(S^{-1}[K^{10}_{0,3} \oplus C_{0,3}] \oplus K^{10}_{0,3} \oplus K^{10}_{0,2}) \oplus \\ &11(S^{-1}[K^{10}_{1,3} \oplus C_{1,3}] \oplus K^{10}_{1,3} \oplus K^{10}_{1,2}) \oplus \\ &13(S^{-1}[K^{10}_{2,1} \oplus C_{2,1}] \oplus K^{10}_{2,3} \oplus K^{10}_{2,2}) \oplus \\ &9(S^{-1}[K^{10}_{3,0} \oplus C_{3,0}] \oplus K^{10}_{3,3} \oplus K^{10}_{3,2})\big] \oplus \\ S^{-1}\big[&14(S^{-1}[K^{10}_{0,3} \oplus C_{0,3}] \oplus K^{10}_{0,3} \oplus K^{10}_{0,2}) \oplus \\ &11(S^{-1}[K^{10}_{1,3} \oplus C_{1,3}] \oplus K^{10}_{1,3} \oplus K^{10}_{1,2}) \oplus \\ &13(S^{-1}[K^{10}_{2,1} \oplus C_{2,1}] \oplus K^{10}_{2,3} \oplus K^{10}_{2,2}) \oplus \\ &9(S^{-1}[K^{10}_{3,0} \oplus C_{3,0}] \oplus K^{10}_{3,3} \oplus K^{10}_{3,2})\big] \oplus \end{aligned} \tag{2.9}$$

$$
\begin{aligned}
p_0 = {} & S^{-1}\big[14(S^{-1}[K^{10}_{0,2} \oplus C_{0,2}] \oplus K^{10}_{0,2} \oplus K^{10}_{0,1})\oplus \\
& 11(S^{-1}[K^{10}_{1,1} \oplus C_{1,1}] \oplus K^{10}_{1,2} \oplus K^{10}_{1,1})\oplus \\
& 13(S^{-1}[K^{10}_{2,0} \oplus C_{2,0}] \oplus K^{10}_{2,2} \oplus K^{10}_{2,1})\oplus \\
& 9(S^{-1}[K^{10}_{3,3} \oplus C_{3,3}] \oplus K^{10}_{3,2} \oplus K^{10}_{3,1})\big]\oplus \\
& S^{-1}\big[14(S^{-1}[K^{10}_{0,2} \oplus C^*_{0,2}] \oplus K^{10}_{0,2} \oplus K^{10}_{0,1})\oplus \\
& 11(S^{-1}[K^{10}_{1,1} \oplus C^*_{1,1}] \oplus K^{10}_{1,2} \oplus K^{10}_{1,1})\oplus \\
& 13(S^{-1}[K^{10}_{2,0} \oplus C^*_{2,0}] \oplus K^{10}_{2,2} \oplus K^{10}_{2,1})\oplus \\
& 9(S^{-1}[K^{10}_{3,3} \oplus C^*_{3,3}] \oplus K^{10}_{3,2} \oplus K^{10}_{3,1})\big]
\end{aligned} \tag{2.10}
$$

$$
\begin{aligned}
3p_0 = {} & S^{-1}\big[14(S^{-1}[K^{10}_{0,1} \oplus C_{0,1}] \oplus K^{10}_{0,1} \oplus K^{10}_{0,0})\oplus \\
& 11(S^{-1}[K^{10}_{1,0} \oplus C_{1,0}] \oplus K^{10}_{1,1} \oplus K^{10}_{1,0})\oplus \\
& 13(S^{-1}[K^{10}_{2,3} \oplus C_{2,3}] \oplus K^{10}_{2,1} \oplus K^{10}_{2,0})\oplus \\
& 9(S^{-1}[K^{10}_{3,2} \oplus C_{3,2}] \oplus K^{10}_{3,1} \oplus K^{10}_{3,0})\big]\oplus \\
& S^{-1}\big[14(S^{-1}[K^{10}_{0,1} \oplus C^*_{0,1}] \oplus K^{10}_{0,1} \oplus K^{10}_{0,0})\oplus \\
& 11(S^{-1}[K^{10}_{1,0} \oplus C^*_{1,0}] \oplus K^{10}_{1,1} \oplus K^{10}_{1,0})\oplus \\
& 13(S^{-1}[K^{10}_{2,3} \oplus C^*_{2,3}] \oplus K^{10}_{2,1} \oplus K^{10}_{2,0})\oplus \\
& 9(S^{-1}[K^{10}_{3,2} \oplus C^*_{3,2}] \oplus K^{10}_{3,1} \oplus K^{10}_{3,0})\big]
\end{aligned} \tag{2.11}
$$

We thus have 4 differential equations and the combined search space of (K^9, C^9, C^{*9}) and p_0 is $2^{32} \cdot 2^8 = 2^{40}$. Therefore, the above 4 equations will reduce this search space of K_{10} to $\frac{2^{40}}{(2^8)^4} = 2^8$. Hence using only one faulty ciphertext one can reduce the search space of AES-128 key to 256 choices. However, the time complexity of the attack is 2^{32} as we have to test all the hypothesis of K^{10} by the above Eqs. (2.8), (2.9), (2.10), and (2.11).

2.2.5.5 Improving the Attack Complexity

In this subsection, we present an improvement to reduce the time complexity of the aforementioned attack to 2^{30} from 2^{32}. Consider the set of equations: (2.8), (2.9), (2.10), and (2.11) in pairs. Observe that each possible pair does not contain all the 16 bytes of the AES key. For example, the pair of equations (2.9) and (2.10) contains 14 key bytes excluding $K^{10}_{0,0}$ and $k^{10}_{0,1}$. This fact can be utilized to reduce the time complexity of the attack. In particular, the list of possible key byte values which is prepared after the first phase of the attack can be split before it is filtered in the second phase of the attack.

Consider the quartet of key-byte values $\{K^{10}_{0,0}, K^{10}_{1,3}, K^{10}_{2,2}, K^{10}_{3,1}\}$ in the first phase of the attack. Let us assume one value of the first quartet is (a_1, b_1, c_1, d_1). As per the property of the S-Box, there will be another value of $K^{10}_{0,0}$, say a_2 which satisfies the system of equation (2.6) with the rest of the key byte values remaining the same. Now, we can divide the list of key-byte values for the quartet $\{K_{0,0}^{10}, K_{1,3}^{10}, K_{2,2}^{10}, K_{3,1}^{10}\}$ into two sub-lists L_1 and L_2 - the first list containing the pair of possible values for the key byte $K_{0,0}^{10}$, and the second list containing the distinct values for the remaining part of the quartet corresponding to these pairs of values for $K_{0,0}^{10}$. This gives us two separate lists of expected size 2^7 as opposed to a single list of expected size 2^8. Similarly, we can create two lists L_3 and L_4 of expected size 2^7 each for the key quartet $\{K_{0,1}^{10}, K_{1,0}^{10}, K_{2,3}^{10}, K_{3,2}^{10}\}$. For the remaining two key quartets, we retain the full lists, say L_5 and L_6, of expected size 2^8 size. We now have six lists instead of four to choose the key byte values from. Hence, we may now divide the key byte filtering phase into two steps as follows:

1. **Step-1**: In this step, we test the keys generated from the first phase of the attack using the pair of equations (2.9) and (2.10). As these equations do not involve the key bytes $K_{0,0}^{10}$ and $K_{0,1}^{10}$, we may ignore the lists L_1 and L_3 in this step. This leaves us with four lists, namely L_2, L_4, L_5 and L_6, with a combined input search space of $2^{2\times7+2\times8} = 2^{30}$. Now, the analysis in this step is expected to reduce this search space from 2^{30} to $2^{30-8} = 2^{22}$ possible combinations.
2. **Step-2**: In the second step, each set of key byte values satisfying Step 1 is tested against the four corresponding combinations of the key bytes $K_{0,0}^{10}$ and $K_{0,1}^{10}$ (two possibilities for each byte), using Eqs. (2.8) and (2.11). As already mentioned, the input search space for this test is 2^{22}, which is further expected to be reduced to $2^{22-2\times8} = 2^6$.

It is now straightforward to see that this approach reduces the critical time complexity for the overall attack from 2^{32} in the previous attack to 2^{30} (in th Step-1 of the filtering phase described above). This gives a speedup of approximately 4 times as compared to the original attack, while still reducing the overall key search space to around 2^8.

Each of the DFA attacks described in this section assume that the fault is restricted to a single byte. It is hopefully clear by now to the reader that restricting the fault onset to a single byte allows a simpler attack analysis, as compared to a scenario where the fault spreads across multiple bytes. At the same time, multiple byte faults are easier to inject in practice. This makes it interesting to see how such faults may be exploited to attack AES, albeit under certain assumptions about the fault locations. This is discussed in the following subsection.

2.2.6 DFA of AES-128 Using Multiple Byte Faults

In this section, we illustrate to the reader the possibility of launching a DFA on AES when the injected fault is not necessarily restricted to a single byte of the cipher state. The assumption behind these attacks is that the bytes in which the faults occur can be grouped under one or more diagonals of the state matrix. The most restrictive scenario assumes that the fault spreads to bytes only within a single diagonal of the state matrix, while the most relaxed scenario allows them to spread across all the

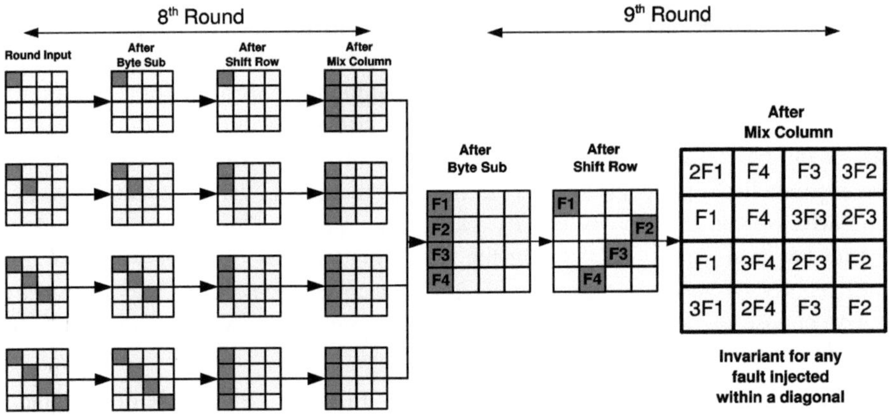

Fig. 2.6 Equivalence of different kinds of faults induced in diagonal D_0 at the input of 8th round of AES

diagonals, that is, anywhere in the state matrix. We now illustrate these scenarios one by one to the reader.

2.2.6.1 Faults Are Restricted to a Single Diagonal

We assume that a fault is injected into the principal diagonal D_0 of the state matrix of AES at the input of the 8th encryption round. Figure 2.6 illustrates different scenarios that could occur under such assumptions. It is interesting to note that irrespective of the number of bytes that are affected by the fault, the subsequent ShiftRow and MixColumn operation ensures that their propagation is eventually confined to the entire single column of the state matrix at the end of the 8th round. This in turn implies that their subsequent propagation across the 9 and 10th rounds of AES is identical. Also, the diagonal number in which the faults are injected directly determines the corresponding column number of the state matrix that is infected at the end of the 8th round.

The byte inter-relations resulting from different affected diagonals are depicted in Fig. 2.7. Note that these inter-relations are independent of the number of affected bytes in each diagonal. The four sets of relations (one for each diagonal) can in turn be used to formulate fault propagation equations. We leave it to the readers as an exercise to figure out that the nature of these equations would be similar to that of Eq. 2.5, and consequently, will reduce the AES final round key search space to 2^{32}. Also, if the adversary is unsure of exactly which diagonal has been affected, she may repeat the analysis four times, once for each diagonal, and retrieve the correct key. This would create a maximum key search space of size $2^{32+2} = 2^{34}$, which can even be brute-forced using modern computational tools.

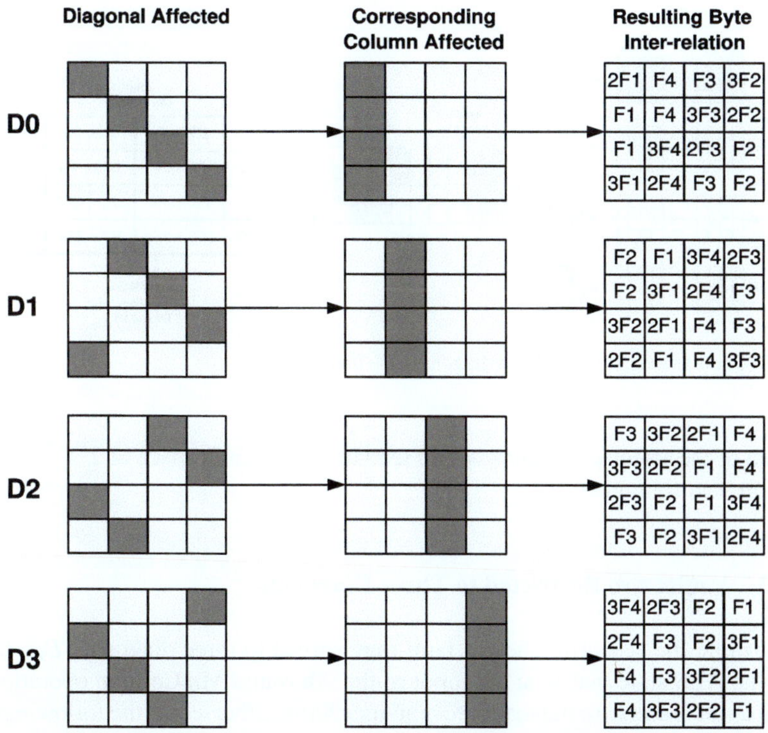

Fig. 2.7 Fault propagation: faults in a single diagonal at the input of 8th round of AES

2.2.6.2 Faults Are Restricted to Two Diagonals

We now consider the scenario where the fault is injected across two diagonals of the AES state matrix at the input of the 9th round MixColumn operation. Consider the fault propagation scenario depicted in Fig. 2.8 where the infected diagonals are D_0 and D_1. Once again, observe that the fault nature at the input and output of the MixColumn operation is an invariant for the number of affected bytes in each of the diagonals. Now, let the faulty values in the first column of the AES state matrix after the 9th round MixColumn operation be denoted as a_0, a_1, a_2 and a_3, respectively. The following inter-relationships hold among these faulty bytes:

$$a_1 + a_3 = a_0$$
$$2a_1 + 3a_3 = 7a_2$$

We can further formulate equations relating the four byte values a_0, a_1, a_2 and a_3, with the correct and faulty ciphertexts C and C^*, respectively, along with the four bytes of the last round key $\{K^{10}_{0,0}, K^{10}_{1,3}, K^{10}_{2,2}, K^{10}_{3,1}\}$. The equations would reduce the key search space for these four bytes to 2^{16} from 2^{32}. Performing a similar analysis

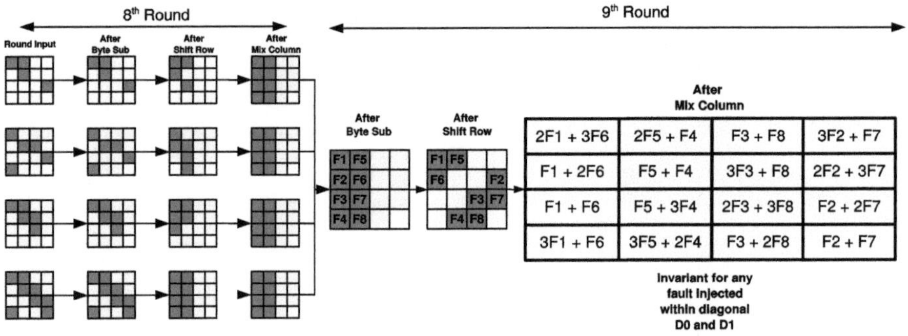

Fig. 2.8 Fault propagation: faults in diagonals D_0 and D_1

for each of the affected columns would lead to a key search space of $2^{16\times4} = 2^{64}$ for the entire last round key.

2.2.6.3 Faults Are Restricted to Three Diagonals

We illustrate the scenario where a fault may spread to three diagonals D_0, D_1 and D_2 of the AES state matrix at the input of the 9th round MixColumn operation. We leave it as an exercise to the reader to convince himself/herself of the following facts:

1. The fault nature at the input and output of the 9th round MixColumn operation is an invariant for the number of affected bytes in each of the diagonals.
2. Let the faulty values in the first column of the AES state matrix after the 9th round MixColumn operation be denoted as a_0, a_1, a_2 and a_3, respectively. The following inter-relationships hold among these faulty bytes:

$$11a_0 + 13a_1 = 9a_2 + 14a_3.$$

3. One can further formulate equations relating the four byte values a_0, a_1, a_2 and a_3, with the correct and faulty ciphertexts C and C^*, respectively, along with the four bytes of the last round key $\{K^{10}_{0,0}, K^{10}_{1,3}, K^{10}_{2,2}, K^{10}_{3,1}\}$. The equations would reduce the key search space for these four bytes to 2^{24} from 2^{32}.
4. Performing a similar analysis for each of the affected columns would lead to a key search space of 2^{96} for the entire last round key.

Finally, it follows from the pattern observed above that if the fault were to affect all four diagonals of the state matrix, the resultant key search space would remain 2^{128}, which is essentially the brute-force key search space and would not lead to an attack (Fig. 2.9).

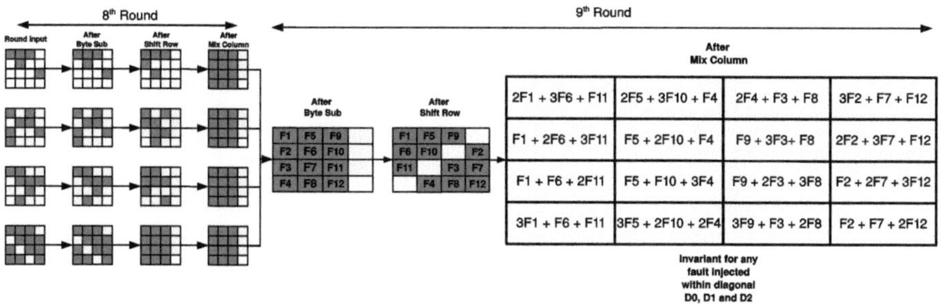

Fig. 2.9 Fault propagation: faults in diagonals D_0, D_1 and D_2

2.2.7 DFA of the AES-128 Key Schedule

All the aforementioned DFA attacks have generally targeted the encryption data path of the AES-128 algorithm. The faults have been injected to corrupt one or more bytes of the state matrix, and the corresponding fault propagation equations have been formulated and solved to recover the secret key. In this final subsection, we present to the readers a DFA attack of a slightly different flavor - namely, a DFA targeting the *key schedule* of AES-128. It is important to note that DFA on AES key schedule leads to more complicated systems of equations as compared to DFA of the encryption data path. This is because a single fault injected in an AES round key at a given location is bound to propagate to further round keys, and consequently corrupt a larger number of state bytes. This leads to a manifold increase in the complexity of the resulting fault propagation relations involving the correct and faulty ciphertext pairs, as well as the corresponding key bytes.

The first complete DFA on AES-128 key schedule was proposed in [45], which required less than 30 pairs of faulty and fault-free ciphertexts to recover the entire key. The attack was further improved in [142] using multi-byte fault models to reduce the required number of faulty and fault-free ciphertext pairs to around 12. An improved attack in [170] recovered the entire last round key sing two pairs of fault-free and faulty ciphertexts and a brute-force search complexity of 2^{48}. Two more attacks on the AES-128 key schedule using two pairs of fault-free and faulty ciphertexts each were proposed in [97, 100]. Finally, [5] presents a highly optimized attack targeting the AES key schedule that requires only one pair of fault-free and faulty ciphertexts to recover the entire key. We present this attack to the readers.

The attack on the key schedule proceeds in two phases. The required faulty ciphertext is generated by inducing a single-byte fault in the first column of the eighth round key while it is being generated. Figure 2.10 demonstrates the fault propagation in the AES key schedule, while Fig. 2.11 demonstrates the fault propagation in the AES state. The first phase of the attack reduces the key search space to 2^{40}, while the second phase reduces the search space further to 2^8. The two phases are discussed below:

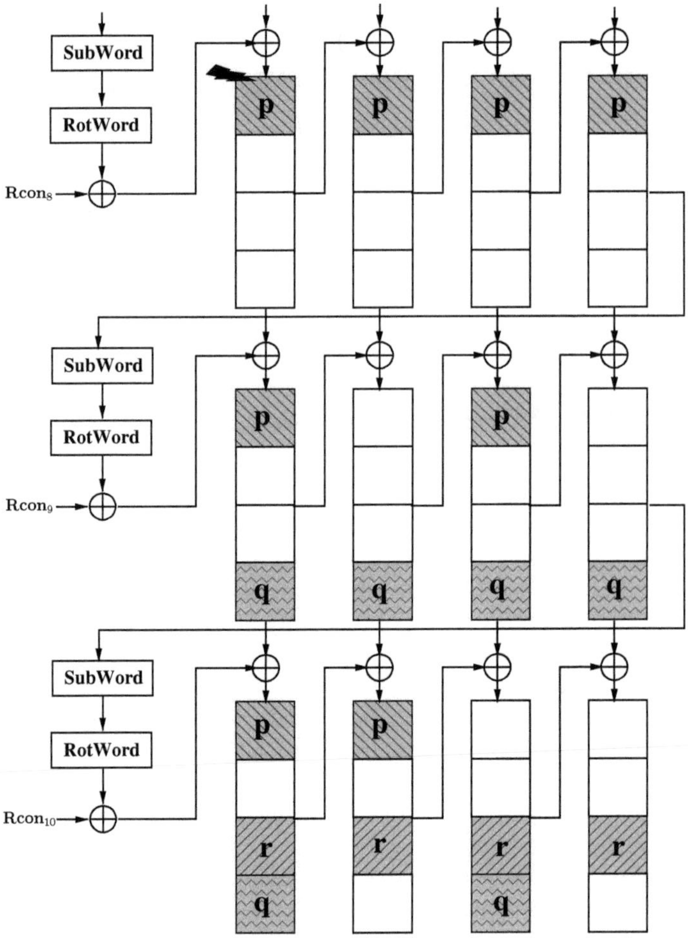

Fig. 2.10 Fault propagation: AES-128 key schedule

- **Phase-1**: As illustrated in Fig. 2.11, the faults injected in the round key K^8 corrupts the first row of the state matrix at the input of the 9th round. After this, the faults are propagated to the entire AES-128 state matrix post the MixColumn operation. Let the fault-free and faulty ciphertext bytes be denoted by C and C^*, respectively. The first column S_2 will produce a set of four differential equations similar to Eq. (2.5) which corresponds to the key quartet $\{K^{10}_{0,0}, K^{10}_{1,3}, K^{10}_{2,2}, K^{10}_{3,1}\}$. Similarly, from other three columns we get three more sets of equations corresponding to the other three key quartets. We denote these four key quartets as K_{q0}, K_{q1}, K_{q2}, and K_{q3} respectively. Note that unlike the DFA on the AES-128 state, we have p, q, and r as extra unknown variables. These three unknown variables are essentially derived from key schedule operation and related by following equations:

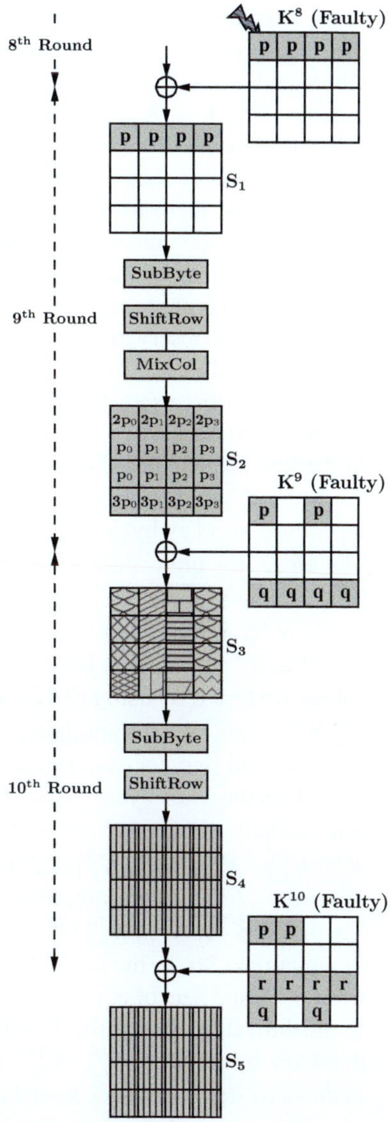

Fig. 2.11 Fault propagation: last 3 rounds of AES-128

$$
\begin{aligned}
q &= S[K^8_{0,3}] \oplus S[K^8_{0,3} \oplus p] \\
&= S[K^9_{0,3} \oplus K^9_{0,2}] \oplus S[K^9_{0,3} \oplus K^9_{0,2} \oplus p] \\
&= S[K^{10}_{0,3} \oplus K^{10}_{0,1}] \oplus S[K^{10}_{0,3} \oplus K^{10}_{0,1} \oplus p]
\end{aligned} \tag{2.12}
$$

$$
\begin{aligned}
r &= S[K^9_{3,3}] \oplus S[K^9_{3,3} \oplus q] \\
&= S[K^{10}_{3,3} \oplus K^{10}_{3,2}] \oplus S[K^{10}_{3,3} \oplus K^{10}_{3,2} \oplus q]
\end{aligned} \tag{2.13}
$$

In the first three sets of equations there are 8 unknown variables (p, q, r, p_i) and the corresponding quartet of key bytes K_{qi}; where i corresponds to the ith quartet. Observe that the fourth set of equations does not contain p. In order to get the quartets K_{q0}, K_{q1}, K_{q2} from the first three sets of equations, we need to test all possible 2^{32} values for (p, q, r, p_i). Additionally, for all possible 2^{32} choices we get 2^{32} hypotheses of each of the quartets. The last set of equation involves only q, r, and p_3; hence, there are only 2^{24} possible hypotheses for K_{q3}.This yields a total of $(2^{32})^3 \cdot 2^{24} = 2^{120}$ possible hypotheses for K^{10} are given by $(2^{32})^3 \cdot 2^{24} = 2^{120}$, which can neither be enumerated nor tested in a practically feasible span of time. To solve this conundrum, a divide-and-conquer technique is applied. The solution is based on the observation that the key bytes $K^{10}_{0,3}$, $K^{10}_{0,1}$, $K^{10}_{3,2}$, $K^{10}_{3,3}$, and (p, q) are also contained in (2.12) and (2.13). These equations can therefore be combined with the last three sets of equation corresponding to the key quartets K_{q1}, K_{q2}, and K_{q3}, respectively.
In the first step we test the possible values of (p, q) For, each of these values we guess the 2^8 values of p_1 in the second set of equations. For each (p, q, p_1) we get the values of 3 key bytes $K^{10}_{0,1}$, $K^{10}_{1,0}$, and $K^{10}_{3,2}$ from the corresponding equations. Therefore, for one value of (p, q) we get 2^8 hypotheses for $(K^{10}_{0,1}, K^{10}_{1,0}, K^{10}_{3,2})$. Similarly, we guess p_3 in fourth set of equations and get 2^8 hypotheses for $(K^{10}_{0,3}, K^{10}_{1,2}, K^{10}_{3,0})$. Therefore, for one hypothesis for (p, q) we get a total of $2^8 \cdot 2^8 = 2^{16}$ hypotheses for 6 key bytes $(K^{10}_{0,1}, K^{10}_{1,0}, K^{10}_{3,2}, K^{10}_{0,3}, K^{10}_{1,2}, K^{10}_{3,0})$. These values are tested by using (2.12), which will reduce the possible values of these 6 key bytes to $\frac{2^{16}}{2^8} = 2^8$ hypotheses.
In the second step, for each hypothesis for the six key bytes, we guess the values of p_2 and get the 3 key bytes $(K^{10}_{0,2}, K^{10}_{1,1}, K^{10}_{3,3})$ from the third set of equations. Therefore, we have a total of $2^8 \cdot 2^8 = 2^{16}$ hypotheses for nine key bytes $(K^{10}_{0,1}, K^{10}_{1,0}, K^{10}_{3,2}, K^{10}_{0,3}, K^{10}_{1,2}, K^{10}_{3,0}, K^{10}_{0,2}, K^{10}_{1,1}, K^{10}_{3,3})$. We use these and get the corresponding values of r from (2.13). Therefore, now using the values of r we can deduce the other 3 key bytes $(K^{10}_{2,3}, K^{10}_{2,0}, K^{10}_{2,1})$ from the corresponding equations in the last three sets of equations. So, in the second step we deduce 2^{16} hypotheses for 12 key bytes from the last 3 sets of equations.
In the third step, we test the 2^8 values for p_0 and get the corresponding choices of the 4 key bytes $\{K^{10}_{0,0}, K^{10}_{1,3}, K^{10}_{2,2}, K^{10}_{3,1}\}$ from the first set of equations. Therefore, in the third step we deduce a total of $2^{16} \cdot 2^8 = 2^{24}$ hypotheses for the 16 key bytes of K^{10} corresponding to one hypothesis for (p, q). Therefore, for all possible 2^{16} hypotheses for (p, q), we will get $2^{24} \cdot 2^{16} = 2^{40}$ hypotheses for K^{40}.

- **Phase-2**: In this phase of the attack we deduce differential equations from the differences in the state matrix S_1 (see Fig. 2.11). In the first row of the state matrix, there exists a 4-byte differential of the form (p, p, p, p). The faulty byte p at the first column of the state matrix is thus as follows:

$$\begin{aligned} p = S^{-1}\big[&14(S^{-1}[K^{10}_{0,0} \oplus C_{0,0}] \oplus K^{9}_{0,0}) \oplus 11(S^{-1}[K^{10}_{1,3} \oplus C_{1,3}] \oplus K^{9}_{1,0}) \oplus \\ &13(S^{-1}[K^{10}_{2,2} \oplus C_{2,2}] \oplus K^{9}_{2,0}) \oplus\ 9(S^{-1}[K^{10}_{3,1} \oplus C_{3,1}] \oplus K^{9}_{3,0})\big] \oplus \\ S^{-1}\big[&14(S^{-1}[K^{10}_{0,0} \oplus C^{*}_{0,0} \oplus p] \oplus (K^{9}_{0,0} \oplus p)) \oplus 11(S^{-1}[K^{10}_{1,3} \oplus C^{*}_{1,3}] \oplus K^{9}_{1,0}) \oplus \\ &13(S^{-1}[K^{10}_{2,2} \oplus C^{*}_{2,2} \oplus r] \oplus K^{9}_{2,0}) \oplus\ 9(S^{-1}[K^{10}_{3,1} \oplus C^{*}_{3,1}] \oplus (K^{9}_{3,0} \oplus q))\big] \end{aligned} \tag{2.14}$$

Similarly, the other three faulty bytes can be expressed by the following:

$$\begin{aligned} p = S^{-1}\big[&14(S^{-1}[K^{10}_{0,1} \oplus C_{0,1}] \oplus K^{9}_{0,1}) \oplus 11(S^{-1}[K^{10}_{1,0} \oplus C_{1,0}] \oplus K^{9}_{1,1}) \oplus \\ &13(S^{-1}[K^{10}_{2,3} \oplus C_{2,3}] \oplus K^{9}_{2,1}) \oplus\ 9(S^{-1}[K^{10}_{3,2} \oplus C_{3,2}] \oplus K^{9}_{3,1})\big] \oplus \\ S^{-1}\big[&14(S^{-1}[K^{10}_{0,1} \oplus C^{*}_{0,1} \oplus p] \oplus (K^{9}_{0,1})) \oplus 11(S^{-1}[K^{10}_{1,0} \oplus C^{*}_{1,0}] \oplus K^{9}_{1,1}) \oplus \\ &13(S^{-1}[K^{10}_{2,3} \oplus C^{*}_{2,3} \oplus r] \oplus K^{9}_{2,1}) \oplus\ 9(S^{-1}[K^{10}_{3,2} \oplus C^{*}_{3,2} \oplus q] \oplus (K^{9}_{3,1} \oplus q))\big] \end{aligned} \tag{2.15}$$

$$\begin{aligned} p = S^{-1}\big[&14(S^{-1}[K^{10}_{0,2} \oplus C_{0,2}] \oplus K^{9}_{0,2}) \oplus 11(S^{-1}[K^{10}_{1,1} \oplus C_{1,1}] \oplus K^{9}_{1,2}) \oplus \\ &13(S^{-1}[K^{10}_{2,0} \oplus C_{2,0}] \oplus K^{9}_{2,2}) \oplus\ 9(S^{-1}[K^{10}_{3,3} \oplus C_{3,3}] \oplus K^{9}_{3,2})\big] \oplus \\ S^{-1}\big[&14(S^{-1}[K^{10}_{0,2} \oplus C^{*}_{0,2}] \oplus (K^{9}_{0,2}) \oplus p) \oplus 11(S^{-1}[K^{10}_{1,1} \oplus C^{*}_{1,1}] \oplus K^{9}_{1,2}) \oplus \\ &13(S^{-1}[K^{10}_{2,0} \oplus C^{*}_{2,0} \oplus r] \oplus K^{9}_{2,2}) \oplus\ 9(S^{-1}[K^{10}_{3,3} \oplus C^{*}_{3,3}] \oplus (K^{9}_{3,2} \oplus q))\big] \end{aligned} \tag{2.16}$$

$$\begin{aligned} p = S^{-1}\big[&14(S^{-1}[K^{10}_{0,3} \oplus C_{0,3}] \oplus K^{9}_{0,3}) \oplus 11(S^{-1}[K^{10}_{1,3} \oplus C_{1,3}] \oplus K^{9}_{1,3}) \oplus \\ &13(S^{-1}[K^{10}_{2,1} \oplus C_{2,1}] \oplus K^{9}_{2,3}) \oplus\ 9(S^{-1}[K^{10}_{3,0} \oplus C_{3,0}] \oplus K^{9}_{3,3})\big] \oplus \\ S^{-1}\big[&14(S^{-1}[K^{10}_{0,3} \oplus C_{0,3}] \oplus (K^{9}_{0,3}) \oplus 11(S^{-1}[K^{10}_{1,3} \oplus C_{1,3}] \oplus K^{9}_{1,3}) \oplus \\ &13(S^{-1}[K^{10}_{2,1} \oplus C_{2,1} \oplus r] \oplus K^{9}_{2,3}) \oplus\ 9(S^{-1}[K^{10}_{3,0} \oplus C_{3,0} \oplus q] \oplus (K^{9}_{3,3} \oplus q))\big] \oplus \end{aligned} \tag{2.17}$$

In the first phase of the attack we have already reduced p, q, r, and K^{10} to 2^{40} choices. Using these values we can get the 9th round fault-free and faulty outputs. As per the attack on the AES-128 key scheduling algorithm (Fig. 2.10), we can directly deduce the 9th round key from the 10th round key. Therefore, for each value of K^{10} we get the corresponding values of K^{9} and can test it using the four equations. There are four equations, and the total search space is 2^{40}. Therefore, the four equations reduce the search space to $\frac{2^{40}}{(2^8)^4} = 2^8$. Hence, in the second phase of the attack we have only 2^8 hypotheses for K^{10}. These can then be used to drive 2^8 hypotheses for the master key.

2.2.7.1 Reducing the Time Complexity of the Attack

The first phase of the attack yields four sets of equations - each corresponding to one of the four key quartets: K_{q0}, K_{q1}, K_{q2}, and K_{q3}. This results in a total search space of 2^{40} possible hypotheses for the entire last round key K^{10}. In the second phase of the attack, each of these hypotheses is tested using a system of four equations, none of which require all 16 bytes of the key. For example, the first equation uses the key bytes $K^{10}_{0,0}$, $K^{10}_{1,3}$, $K^{10}_{2,2}$, $K^{10}_{3,1}$, and nine more key bytes corresponding to the four ninth round key bytes $K^9_{0,0}$, $K^9_{1,0}$, $K^9_{2,0}$, $K^9_{3,0}$. This leads to a total requirement of 13 key bytes. The remaining equations require 10 key bytes each. The first phase of the attack uses (2.12) and (2.13) owing to their dependencies on the key bytes $K^{10}_{0,3}$, $K^{10}_{0,1}$, and $K^{10}_{3,3}$, $K^{10}_{3,2}$.

In the second phase, the adversary tests one equation at a time. It is logical to start with equations involving fewer key bytes; for example, the third equation only requires 11 bytes of K^{10} (ten key bytes plus one for $K^{10}_{0,3}$ since it depends on $K^{10}_{0,1}$ in (2.12)). The key hypotheses that satisfy this equation are retained and combined with five other byte hypotheses. These combined solutions are subsequently tested using the remaining equations. Now, observe that the third equation requires $2^{40}/2^5 = 2^{35}$ possible unique choices over the 11 key bytes. These choices are now combined with a maximum of 2^5 possible hypotheses for the remaining key bytes, and tested using the four equations. Each of these equations reduces the number of candidate hypotheses by 2^8, leaving $\frac{2^{40}}{(2^8)^4} = 2^8$. Quite evidently, this search space is achieved using an attack time complexity of 2^{35}.

2.2.8 DFA of AES-192 and AES-256

In the case study, we have principally focused on DFA of AES-128, which the most popular and widely adopted version of the AES block cipher in both literature as well as practice. There exist DFA attacks on other versions of AES, namely AES-192 and AES-256 as well. A major difference of AES-192 and AES-256 from AES-128 is that the last round key in either of these two variants is not sufficient to uniquely recover the entire master secret key. Hence, an adversary must retrieve two consecutive round keys - for the last round as well as the penultimate round, to obtain the master secret key. Several such DFA propositions have been put forth in the literature. The proposition in [107] is based on two different fault models which requires 6 and 3000 pairs of fault-free and faulty ciphertexts. The proposition in [169] exploits the relations between the round keys of the key scheduling algorithm, and requires three pairs of correct and faulty ciphertexts for AES-192, and two pairs of correct and faulty ciphertexts and two pairs of correct and faulty plaintexts for AES-256. The attack was further improved in [96], with the DFA on AES-192 requiring two pairs of fault-free and faulty ciphertexts, and the DFA on AES-256 requiring three pairs of fault-free and faulty ciphertexts. The best known DFA of AES-256 was proposed in [6], which requires two pairs of fault-free and faulty ciphertexts, with a brute-force search complexity of 2^{16} and an overall attack complexity of 2^{32}. Attacks proposed in [132] have also targeted the respective key-scheduling algorithms for AES-192 and AES-256, and require two and three pairs of fault-free and faulty ciphertexts, respectively.

2.3 Differential Fault Analysis of Stream Ciphers

In this section, we present to the reader DFA attacks on a different class of symmetric-key cryptographic primitives - stream ciphers. In a stream cipher, a stream of plaintext data is combined with a pseudo-random stream of key bits, generated using a keystream generation algorithm. Unlike block ciphers that operates on a chunk of plaintext data with pre-defined width, stream ciphers can operate more granularly on chunks as small as bytes of data. Similar to block ciphers, stream ciphers have also been subjected to DFA [14, 15, 28, 88, 161]. Unlike block ciphers, DFA on stream ciphers usually target the state registers in the keystream generation module. The first practical demonstration of DFA on stream ciphers was presented in [63]. The authors targeted Grain-128 [77], the eSTREAM finalist, using low-cost fault clock glitch-based fault injection setup. The fault model used is single-bit flips, occurring in the feedback path of the Feedback Shift Register, which also happens to be the critical path in the circuit as per their design. The fault is subsequently propagated to multiple contiguous positions in the state register, which is then exploited by the DFA algorithm to recover the key.

2.3.1 A Brief Description of Grain-128

We present a brief description of Grain-128 here for the ease of understanding of the readers. A more detailed description can be found in [77]. The internal state IS_i of the key-generation algorithm of Grain-128 consists of two feedback shift registers X (non-linear) and Y (linear) with inner states $X_i = (x_i, \ldots, x_{i+127})$ and $Y_i = (y_i, \ldots, y_{i+127})$ respectively, at the beginning of the Pseudo-Random keystream Generation Algorithm (PRGA) round i (≥ 1),

$$IS_i = (\underbrace{x_i, \ldots, x_{i+127}}_{X_i} \mid \underbrace{y_i, \ldots, y_{i+127}}_{Y_i}).$$

The secret *key* $(k_1, \ldots, k_{128})$ and *IV* $(IV_1, \ldots, IV_{96})$ are used to initialize the inner state as follows:

$$(\underbrace{k_1, \ldots, k_{128}}_{X} \mid \underbrace{IV_1, \ldots, IV_{96}, 1, \ldots, 1}_{Y}).$$

The keystream bit z_i and the new inner state bits x_{i+128}, y_{i+128} of Grain-128 registers X, Y are generated respectively as follows:

$z_i = h(X_i, Y_i)$, $x_{i+128} = u(X_i, Y_i)$, $y_{i+128} = v(Y_i)$ where,

$h(X_i, Y_i) = x_{i+2} + x_{i+15} + x_{i+36} + x_{i+45} + x_{i+64} + x_{i+73} + x_{i+89} + y_{i+93} +$
$x_{i+12}x_{i+95}y_{i+95} + x_{i+12}y_{i+8} + y_{i+13}y_{i+20} + x_{i+95}y_{i+42} + y_{i+60}y_{i+79}$,

$u(X_i, Y_i) = y_i + x_i + x_{i+26} + x_{i+56} + x_{i+91} + x_{i+96} + x_{i+3}x_{i+67} + x_{i+11}x_{i+13} +$
$x_{i+17}x_{i+18} + x_{i+27}x_{i+59} + x_{i+40}x_{i+48} + x_{i+61}x_{i+65} + x_{i+68}x_{i+84}$,

$v(Y_i) = y_i + y_{i+7} + y_{i+38} + y_{i+70} + y_{i+81} + y_{i+96}$.

2.3.2 The DFA Attack Model

The attack proposed in [63] assumes that the actual cipher device can be re-keyed with the same *key-IV* before each fault injection trial. For any two integers a and b, with $a \leq b$, we denote the set $\{x : x \text{ is an integer with } a \leq x \leq b\}$ simply by $[a, b]$. Also if $V = (V_1, \ldots, V_p)$ is a vector of length p, we denote V_i simply by $V(i)$ for all $i \in [1, p]$. We shall also use the following notations:

- For any integer i, $\emptyset + i = \emptyset$ ($\emptyset$ being the empty set).
- For any set S of integers and for any integer i, $S + i = \{s + i : s \in S\}$.
- For any set S if $s \in S$ implies that s is a set of integers then for any integer i, $S + i = \{s + i : s \in S\}$.

2.3.2.1 Fault Location

If a fault injection trial flips exactly the bits at the r distinct positions given by $\phi = \{\phi_1, \ldots, \phi_r\}$ of the internal state, only at the PRGA round t, then the set ϕ will be called a *fault position* and the ordered pair (ϕ, t) will be called a *fault location* or simply a *fault*.

Remark The authors of [63] consider faults at a single PRGA round or multi-round faults equivalent to a fault at a single PRGA round. Let Γ be the set of all possible fault positions corresponding to a fault model Σ. We shall represent Σ simply by Γ.

2.3.2.2 The XOR Differential Keystream

Let IS_i be the internal state of the cipher at PRGA round i $(i \geq 1)$. Let us consider a fault (ϕ, t). Let $IS_i^{\phi,t}$ be the faulty internal state and let $z_i^{\phi,t}$ be the faulty output key bit at that PRGA round i. Then $d_i^{\phi,t} = z_i^{\phi,t} + z_i$ is the XOR difference of the normal (fault free) keystream bit z_i from the faulty one $z_i^{\phi,t}$ at the PRGA round i.
For given n we denote, $d^{\phi,t,n} = (d_1^{\phi,t}, \ldots, d_n^{\phi,t})$.

Remark One should note that each $d_i^{\phi,t}$ may be thought of as a function of the *Key-IV* pair. Since the cipher device is re-keyed before each fault injection, after the fault injection, at the fault injection PRGA round t we have, $IS_t^{\phi,t}(e) = IS_t(e) + 1$, $\forall e \in \phi$ and $IS_t^{\phi,t}(e) = IS_t(e)$, $\forall e \in [1, 256] \setminus \phi$.

2.3.2.3 Signature of Fault Locations

After a fault (ϕ, t) is injected in the PRGA round t, *we shall study the l PRGA rounds* $t, \ldots, t + l - 1$. We consider the XOR differential keystream $d^{\phi,t,n} = (d_1^{\phi,t}, \ldots, d_n^{\phi,t})$

where $n = t + l - 1$. It should be noted that $d_1^{\phi,t} = \cdots = d_{t-1}^{\phi,t} = 0$ as the fault is injected at the PRGA round t. We now treat *key-IV* as variables and each $d_i^{\phi,t} : \mathrm{GF}(2)^{128+96} \rightarrow \mathrm{GF}(2)$ as a function of the *Key (128 bit) - IV (96 bit)* pair. For simplicity we write $d_i^{\phi,t} \notin \{0, 1\}$, to mean that $d_i^{\phi,t}$ is not a constant function.

For certain (ϕ, t) there may be some special values in $d^{\phi,t,n}$ such that:

1. $d_i^{\phi,t} = b, b \in \{0, 1\}$ irrespective of *Key-IV*.
2. $d_i^{\phi,t}, d_j^{\phi,t} \notin \{0, 1\}, i \neq j$, but $d_i^{\phi,t} = d_j^{\phi,t}$ happens deterministically irrespective of *Key-IV*.
3. $d_i^{\phi,t}, d_j^{\phi,t} \notin \{0, 1\}, i \neq j$, but $d_i^{\phi,t} = d_j^{\phi,t} + 1$ happens deterministically irrespective of *Key-IV*.

For each fault location (ϕ, t) the signature [161], $sig_{\phi,t}$ of the fault (ϕ, t) is defined to be the 4-tuple $sig_{\phi,t} = (sig^1_{\phi,t}, sig^0_{\phi,t}, sig^=_{\phi,t}, sig^{\neq}_{\phi,t})$ as explained below, where each of $sig^e_{\phi,t}$, $e \in \{1, 0, =, \neq\}$ will be called a component of $sig_{\phi,t}$. We now define the following:

$$sig^b_{\phi,t} \overset{\text{def}}{=} \{i \in [t, n] : d_i^{\phi,t} = b\},\ b \in \{0, 1\}$$

$$sig^=_{\phi,t} \overset{\text{def}}{=} \{\{i_1, \ldots, i_p\} : i_1, \ldots, i_p \in [t, n],\ d_{i_1}^{\phi,t} = \cdots = d_{i_p}^{\phi,t} \notin \{0, 1\}$$

and $\exists$ no $i_r \in [t, n]$ other than $i_1, \ldots, i_p$ such that $d_{i_r}^{\phi,t} = d_{i_1}^{\phi,t}\}$.

$$sig^{\neq}_{\phi,t} \overset{\text{def}}{=} \{\{i, j\} : i, j \in [t, n], d_i^{\phi,t} + d_j^{\phi,t} = 1 \text{ and } d_i^{\phi,t}, d_j^{\phi,t} \notin \{0, 1\}\}$$

If $d_i^{\phi,t} = b,\ b \in \{0, 1\}$ holds irrespective of *Key-IV*, we shall say that: "all the XOR differential keystreams are fixed to b at the position i under the fault (ϕ, t)". One should note the following:

1. $sig^b_{\phi,t}$ is the set of the positions (PRGA rounds) where the XOR differential keystreams are fixed to b under the fault (ϕ, t).
2. $sig^=_{\phi,t}$ gives sets of PRGA rounds where the XOR differential keystreams are deterministically equal (but not fixed) irrespective of *Key-IV* under the fault (ϕ, t).
3. $sig^{\neq}_{\phi,t}$ gives pairs of PRGA rounds where the XOR differential keystreams are deterministically different (but not fixed) irrespective of *Key-IV* under the fault (ϕ, t).

Remark The signatures are *Key-IV* independent and depend only on fault locations and the cipher design. Since signature of faults are constructed for finitely many PRGA rounds it may happen that (1) for some fault locations some signature component becomes completely void and (2) signatures of two fault locations match exactly with each other. We now use the following notation: For any integer i, we have:

$$sig_{\phi,t} + i \overset{\text{def}}{=} (sig^1_{\phi,t} + i, sig^0_{\phi,t} + i, sig^{=}_{\phi,t} + i, sig^{\neq}_{\phi,t} + i)$$

Theorem 1 *For any fault* (ϕ, t), $sig_{\phi,t} = sig_{\phi,1} + t - 1$.

Proof Let fault be injected at the PRGA round 1 at position ϕ and $(s_1, s_2, \ldots, s_{256})$ be the corresponding internal state (IS) where each of s_j is a variable. We now assume that the XOR differential keystream bit $d_i^{\phi,1} = 1$. Then this happens (at the PRGA round i) independent of the internal state at the PRGA round 1. Thus $d_i^{\phi,1} = 1 \Leftrightarrow d_{t+i-1}^{\phi,t} = 1$ and hence $i \in sig^1_{\phi,1} \Leftrightarrow t + i - 1 \in sig^1_{\phi,t}$. This shows that $sig^1_{\phi,t} = sig^1_{\phi,1} + t - 1$. With similar arguments the theorem follows.

Remark **(1)** In consequence of the above theorem it can be said that if a fault is injected in the same position then a pattern is generated from the fault injection PRGA round, in the XOR differential keystream, all previous keystream bits being 0's.

Remark **(2)** For any fault (ϕ, t) if $t = 1$, we shall drop the subscript 't' form its signature and signature components. Thus with this simplified notation $sig_{\phi,1} = (sig^1_{\phi,1}, sig^0_{\phi,1}, sig^{=}_{\phi,1}, sig^{\neq}_{\phi,1})$ becomes $sig_{\phi} = (sig^1_{\phi}, sig^0_{\phi}, sig^{=}_{\phi}, sig^{\neq}_{\phi})$ and in this case sig_{ϕ} will be called the signature of the fault position ϕ.

We now present methods for computing the signature components of sig_{ϕ}.

2.3.3 Signature Generation

Let us consider a fault position ϕ. We generate sig^1_{ϕ}, sig^0_{ϕ} using the probabilistic algorithm $GenSig10$ and $sig^{=}_{\phi}$, $sig^{\neq}_{\phi}$ using the deterministic algorithm $GenSigSym$ as described below. The deterministic algorithm $GenSigSym$ has implementation limitations. However, the probabilistic algorithm $GenSig10$ is much more efficient.

Algorithm 1 $GenSig10$

Input: ϕ, Ω, L_1
Output: $sig^0_{\phi}, sig^1_{\phi}$
1: **for** **do**Ω number of distinct uniformly random independent *Key-IV*:
2: Generate XOR differential keystreams upto round L_1 under the fault $(\phi, 1)$
3: **end for**
4: Find positions (PRGA rounds) at which all the generated XOR differential keystreams are fixed
5: If $b \in \{0, 1\}$ is at a fixed position : then add the position to sig^b_{ϕ}
6: **return** $sig^0_{\phi}, sig^1_{\phi}$

Algorithm 2 $GenSigSym$

Input: ϕ, L_2 (the number of PRGA rounds used by the keystream generation algorithm)
Output: $sig_{\phi}^{=}, sig_{\phi}^{\neq}$
1: Define 256 symbolic variables over $GF(2)$ and initialize the inner state with these symbolic variables. This will represent the inner state at the beginning of the PRGA round 1.
2: Compute symbolically the XOR differential keystream $d^{\phi,1,L_2}$
3: Observe the XOR differential keystream and compute $sig_{\phi}^{=}, sig_{\phi}^{\neq}$
4: **return** $sig_{\phi}^{=}, sig_{\phi}^{\neq}$

Let i be an actual fixed position (PRGA round) for $b \in \{0, 1\}$ under the fault $(\phi, 1)$. Then the algorithm $GenSig10$ will surely append i to sig_{ϕ}^{b}. But if i is not a fixed position, then Pr (The algorithm appends i to sig_{ϕ}^{b}) $= 1/2^{\Omega}$ provided we assume that 0 and 1 are equally probable at the position i (since i is not a fixed position) and the XOR differential bit generated at the position i for each *Key-IV* pair are independent (*Key-IV* pairs are distinct uniformly random and independent). Therefore taking large value of Ω it can be guaranteed that the algorithm generates correct signatures with very small failure probability. e.g., simply taking $\Omega = 1000$ we have $1/2^{\Omega} = 10^{-\Omega log_{10} 2} \approx 10^{-301}$ which is practically negligible.

Remark The algorithm $GenSig10$ is generic in nature and is capable of coping with any computationally feasible fault model Γ. It taps statistical weakness of the cipher under DFA.

2.3.4 Fault Location Determination

Let Γ be a fault model and the adversary has computed sig_{ϕ} for all $\phi \in \Gamma$. Adversary is fully confident that any injected fault (position) will be in Γ. In this stage the adversary actually injects a fault into the cipher device and compares the XOR differential keystream with pre-computed signatures of all possible faults in order to identify the fault location. We define, $n_1 = max(\bigcup_{\phi \in \Gamma}(sig_{\phi}^{1} \bigcup sig_{\phi}^{0}))$, $n_2 = max(\bigcup_{\phi \in \Gamma}(\bigcup_{A \in sig_{\phi}^{=} \bigcup sig_{\phi}^{\neq}} A))$ and $end(\Gamma) = max(n_1, n_2)$. The steps for fault location determination are described next.

1. Obtain the fault-free keystream. In this stage we need an XOR differential keystream of length $n = end(\Gamma) + T - 1$ to match it with all possible pre-computed signatures, if fault is injected at the known PRGA round T.
2. Let a fault be injected at an unknown position ψ at the known PRGA round T. Compute the faulty keystream and obtain $d^{\psi,T,n} = (d_1^{\psi,T}, \ldots, d_n^{\psi,T})$.
3. $pf = allPossibleFaults\_known(\Gamma, T, d^{\psi,T,n})$ as described in Algorithm 4.

The basic idea is to check whether the pre-computed pattern (signature) of a fault location occurs in the XOR differential keystream. If the pattern due to a fault location (ϕ, t) occurs in the XOR differential keystream, then it is a possible fault location, otherwise we reject it.

Algorithm 3 *isaPossibleFault*

Input: $\phi, t, d^{\psi,t,n}$
if $sig^1_{\phi,t} \subseteq support^1$ **then**
  if $sig^0_{\phi,t} \subseteq support^0$ **then**
    if $sig^{=}_{\phi,t} == \emptyset$ **then**
      if $sig^{\neq}_{\phi,t} == \emptyset$ **then**
        return True
      else if $\forall\{i,j\} \in sig^{\neq}_{\phi,t} \Rightarrow d^{\psi,t}_i + d^{\psi,t}_j = 1$ **then**
        return True
      else
        return False
      end if
    else if $\forall A \in sig^{=}_{\phi,t}$ and $\forall i, j \in A \Rightarrow d^{\psi,t}_i = d^{\psi,t}_j$ **then**
      if $sig^{\neq}_{\phi,t} == \emptyset$ **then**
        return True
      else if $\forall\{i,j\} \in sig^{\neq}_{\phi,t} \Rightarrow d^{\psi,t}_i + d^{\psi,t}_j = 1$ **then**
        return True
      else
        return False
      end if
    else
      return False
    end if
  else
    return False
  end if
else
  return False
end if

Algorithm 4 *allPossibleFaults_known*

Input: $\Gamma, t, d^{\psi,t,n}$
Output: pf (the set of all possible faults)
$pf = \emptyset$
for each $\phi \in \Gamma$ **do**
  if $isaPossibleFault(\phi, t, d^{\psi,t,n})$ **then**
    $pf = pf \bigcup \{(\phi, t)\}$
  end if
end for
return pf

It should be noted that from the construction it immediately follows that the actual fault location $(\psi, T) \in pf$. Now if pf is a singleton then, (ψ, T) is uniquely determined. When pf is not singleton we do not need to reject the case as a failure. We shall address the issue in the next section.

2.3.5 *Recovering the Internal State of the Cipher*

The adversary wishes to recover the internal state of the cipher at the PGRA round t. In the online phase, the adversary inject faults to the cipher device at PRGA round t, re-keying each time, for m times. Let Q_j be the set of possible faults returned by the fault location determination algorithm at the jth fault injection trial, $z^j = (z_1^j, \ldots, z_n^j)$ being the corresponding faulty keystream of length n $(> t)$, $\forall j \in [1, m]$. We now consider the Cartesian product set $Q = Q_1 \times \cdots \times Q_m$. Then one of the elements, say α of Q corresponds to the actual m injected faults. The adversary starts with the following information:

- m fault locations $\beta = ((\phi_1, t), \ldots, (\phi_m, t)) \in Q$.
- fault free keystream $z = (z_1, \ldots, z_n)$ of length n.
- m faulty keystreams $z^1, \ldots, z^m$ each of length n.

2.3.5.1 Generating Polynomial Equations

We use procedure similar to [161] in order to obtain a system of polynomial equations, modifying the *fault injection strategy* in order to cope with multi-bit faults at any targeted round.

Let the fault free internal state at the PRGA round i $(\geq t)$ be $IS_i = (X_i, Y_i)$ where $X_i = (x_i, \ldots, x_{i+127})$ and $Y_i = (y_i, \ldots, y_{i+127})$, the internal state at PRGA round t being $IS_t = (x_t, \ldots, x_{t+127}, y_t, \ldots, y_{t+127})$. We treat each x_i and y_i as variables and consider the PRGA rounds $t, \ldots, n$. Corresponding to each key-stream bit z_i, we introduce two new variables x_{i+128}, y_{i+128} $(i \geq t)$ and obtain the following three equations: $z_i = h(X_i, Y_i), x_{i+128} = u(X_i, Y_i), y_{i+128} = v(Y_i)$. Thus we have in total $2N + 256$ variables and $3N$ equations, where $N = n - t + 1$.

Let us now consider the fault (ϕ_j, t). Since the cipher device is re-keyed before each fault injection, after the fault injection, if the faulty internal state at PRGA round i be IS_i^j then at the targeted fault injection PRGA round t we have, $IS_t^j(e) = IS_t(e) + 1$, $\forall e \in \phi_j$ and $IS_t^j(e) = IS_t(e)$, $\forall e \in [1, 256] \setminus \phi_j$. Again corresponding to each key-stream bit z_i, we introduce two new variables x_{i+128}^j, y_{i+128}^j $(i \geq t)$ respectively for NLFSR and LFSR and obtain three more equations. In this case we have additional $2N$ variables and $3N$ equations.

Thus if we consider m faults, after these many re-keyings, the total number of variables is $2(m + 1)N + 256$ and the total number of equations is $3(m + 1)N$.

Now the system of polynomial equations are simply passed on to the SAT solver in sage for extracting solution for the variables $x_t, \ldots, x_{t+127}, y_t, \ldots, y_{t+127}$ and to mean this we shall use the phrase that "$(\beta, t, n, z, z^1, \ldots, z^m)$ *are passed on to the SAT solver*".

2.3.5.2 Recovering the Internal State with SAT Solver

Now we pass $(\beta, t, n, z, z^1, \ldots, z^m)$ to the SAT solver, for each element β of Q. Multiple solutions may be obtained. Solution from the element α, if returned, will surely correspond to the actual internal state of the cipher at the PRGA round t. Assuming each returned solution as a possible internal state at the PRGA round t, we simply use "Guess and Determine Strategy" [157] to detect the correct internal state. If we have a match, the internal state together with the actual fault locations (not needed any more) will be recovered.

If the cardinality of Q and SAT solving time for m faults are low then the internal state can be recovered in reasonable time with 100% success. Otherwise we have to re-key the cipher device for more fault injection trials.

2.3.6 Experimental Results: Faults Generated in Hardware

2.3.6.1 Fault Injection Setup

In this section, we present the experimental results of fault injection by *clock glitching* on a hardware implementation of *Grain-128*. To the best of our knowledge, we report for the very first time the actual chip results for a fault attack on any stream cipher. An input clock was provided to a *Grain-128* Spartan-3A (XC3S400A) FPGA implementation from an external function generator. A fast clock of 20 times the frequency compared to the input slow clock was used to introduce a clock glitch at a fixed PRGA round. The fast clock was derived from the slow clock using a Xilinx Digital Clock Manager (DCM) module in the design and the states of the registers were monitored using Chipscope Pro 12.3 analyzer. We obtained the correct ciphertext (corresponding to the fault free *Grain-128* internal state) for input slower clock frequencies up to 7 MHz or fast clock frequencies up to 140 MHz. We gradually increased the input slower clock frequency in steps of 0.1 MHz and captured the corresponding states of the registers at each step. The number of attempts to inject a fault at each step was 1024. In Fig. 2.12, we plot the nature of the induced faults with respect to the frequency of the fast clock for different *key-IV* pairs.

The faults observed were all single bit ones and in each case the faults affected only the 128th bit of the NFSR ($bitNLFSR_{128}$) due to clock glitch introduced set up time violations. This is because the critical path of *Grain-128* is through the NLFSR feedback [117]. From the experimental results we conclude that the single bit faults (if injected) were biased at a particular bit position irrespective of the initial register states. However, the frequency of occurrences of single bit faults varied for different *key-IV* pair initializations of the cipher. The reason for this variation may be due to the data dependent nature of fault sensitivity [111].

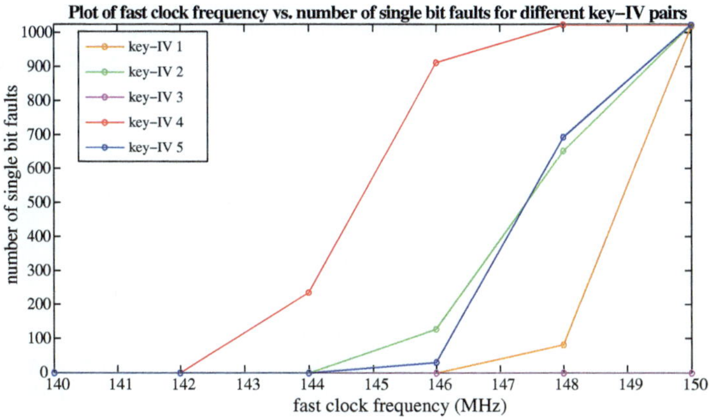

Fig. 2.12 Plot of fast clock frequency versus number of single bit faults for different *key-IV* pairs

2.3.6.2 Introducing Multi-Bit Faults

The reported DFAs on Grain 128 requires at least 3 distinct single bit faults [161]. However, as we see using fault induction techniques like clock glitches, only a single bit is affected (namely due to the critical path), and thus we do not get the required number of faulty outputs to perform the analysis. One may employ more costly injection methods, however in such case also there are chances of the faults being spread to multiple bits. We try to tweak the fault induction technique using clock glitches to obtain distinct faults which can be exploited by the attack developed in this chapter. The experimental results of our clock glitch set up revealed that targeting a particular PRGA round leads to only single bit faults at $bitNLFSR_{128}$ with a very high bias. Therefore, in order to induce multi-bit faults in cipher's internal state within k-neighborhood bits, an adversary can introduce clock glitches in k consecutive PRGA rounds provided the targeted bits lie between the two bounding tap positions of the shift register. According to this modified multi-round fault injection technique, if a fault occurred in the first targeted PRGA round at $bitNLFSR_{128}$, it will propagate (shifted) to $bitNLFSR_{127}$ in the next PRGA round and so on. Therefore, there is a chance of fault injection up to k adjacent bits in the final targeted PRGA round (i.e. the kth round). In Table 2.1, we present the experimental results of such multi-round fault injection (considering $k = 5$) for a randomly chosen *key-IV* pair.

2.3.6.3 DFA of Grain-128: Based on Simulations

Attack strategy in this chapter is generic. In particular we demonstrate the attack strategy for the scenario, denoted by the symbol nbdMBF^k, in which randomly chosen at most k consecutive location (k-neighborhood bit fault) can be disturbed by a single fault injection without knowing the locations or the exact number of bits

Table 2.1 *Grain-128* fault distribution pattern versus fast clock frequency

Fast clock frequency (MHz)	No fault	Single bit fault	Multi-bit fault
130	1024	0	0
140	1024	0	0
150	1024	0	0
160	1024	0	0
170	1024	0	0
180	1024	0	0
190	0	$bitNLFSR_{128}$ (1024)	0
200	0	$bitNLFSR_{125}$ (2)	$bitNLFSR_{128}$ (1022), $bitNLFSR_{125}$ (1022)
210	0	0	$bitNLFSR_{127}$ (1024), $bitNLFSR_{126}$ (1024), $bitNLFSR_{125}$ (1024)
220	0	0	$bitNLFSR_{127}$ (1024), $bitNLFSR_{126}$ (1024), $bitNLFSR_{125}$ (1024), $bitNLFSR_{124}$ (1024)

the injected fault has altered. In this chapter we consider the popular convention of treating $IS(e), IS(e+1), IS(e+2), \ldots, IS(e+p-1)$ as p neighboring bits (IS representing the internal state of the cipher) but in real life the arrangement may not follow this pattern. However this does not affect our analysis.

Faults in the Model nbdMBF^k. let Γ_k be set of all possible fault positions in this case. In the case when exactly r bits are flipped, the fault positions are of the form $i, i+j_1, i+j_2, \ldots, i+j_{r-1}$ where $i \in [1, 256-j_{r-1}]$ and $(j_1, j_2, \ldots, j_{r-1})$ follows lexicographic ordering (increasing) without repetition of length $r-1$ in the range $[1, k-1]$. e.g., when $k=4, r=3$ possible (j_1, j_2) are $(1,2), (1,3), (2,3)$ and hence the fault positions are of the form $(i, i+1, i+2)$, $(i, i+1, i+3)$, $(i, i+2, i+3)$.

i	$i+1$	$i+2$	$i+3$
*	*	*	
*	*		*
*		*	*

i	$i+1$	$i+2$	$i+3$
i	$i+1$	$i+2$	
i	$i+1$		$i+3$
i		$i+2$	$i+3$

j_1	j_2
1	2
1	3
2	3

For each fault position ϕ we used the algorithm $GenSig10(\phi, \Omega, L_1)$ to generate sig_ϕ^1 and sig_ϕ^0 by taking $L_1 = 1000$, $\Omega = 1000$ while $GenSigSym(\phi, L_2)$ was used to generate $sig_\phi^=$ and $sig_\phi^{\neq}$ by taking $L_2 = 150$. Here we present experimental results for (comparing) $k = 1, 2, 3, 4, 5$. The case for $k = 1$ results in the same fault model considered in [161].

Our Arsenal

1. One standalone desktop PC with AMD 4.0 GHz FX-8350 processor and 32 GB RAM, referred to as AMD.
2. A Beowulf Cluster of 20 desktop PC each with 2.60 GHz Intel Pentium E5300 Dual-core processor and 4 GB RAM connected via LAN and set up for Distributed Parallel Computing, referred to as BEOWULF.
3. SAT solver Cryptominisat - 2.9.6 installed in SAGE - 6.1.1.

The BEOWULF cluster (all available cores) was used to (1) generate sig^1_ϕ and sig^0_ϕ, (2) compute the success rates (probabilities) and (3) SAT solving where as the standalone AMD (only a single core) was used to generate $sig^=_\phi$ and $sig^{\neq}_\phi$.

All simulation based experiments were performed assuming PRGA round 1 as the target round.

2.3.7 *Probability of Identifying a Random Fault Location*

For each column of the following table, we considered a set of 2^{20} experiments.

Grain-128					
k	1	2	3	4	5
POS	1.0	1.0	0.997974	0.958305	0.927482
Avg Fault	1.0	1.0	1.000051	1.040921	1.107610

Explanation: For $k = 5$, POS (Avg Fault) = 0.927482 (1.107610) means that the fault location determination algorithm had uniquely identified actual fault locations with a probability of 0.927482 and average number of faults returned by the fault location determination algorithm in 2^{20} experiments is 1.107610 which is very low. For $k = 3, 4$ the average number of faults returned by the fault location determination algorithm are even lower. For $k = 1, 2$ the success rates are 100% in uniquely identifying actual fault locations.

2.3.7.1 SAT Solving: Results

Behavior of SAT solvers and the time to return a solution could hardly be predicted. For each SAT solving trial (with cutting number 4) we first generated an inner state (by choosing key-IV randomly) and m random faults uniformly and independently. For each SAT solving trial (compiled codes were not used) we allocated a time limit of 4 hours. If the SAT solver does not self terminate within that allocated time we agree to terminate it forcefully and mark the case as a TIMEOUT (which may have resulted in a success if enough time was permitted). Very few such case occurred for

$m = 4$ during experimentation while the case for $m = 5$ always resulted in a success. For each row of the following table, we performed SAT experiments independently.

Abbreviations: (POS, Probability of Success), (NOE, Number of Experiments), (NOT, Number of Timed out cases).

Grain-128					Time in Seconds if SUCCESS		
k	m	N	NOE	NOT	MinTime	MaxTime	AvgTime
1	4	256	160	5	9.80	12301.04	650.77
2	4	256	160	4	9.38	10803.83	811.91
3	4	256	160	7	10.40	14071.91	426.38
4	4	256	160	6	10.25	14378.43	679.92
5	4	256	160	8	8.88	10294.13	419.93
1	5	256	1500	0	5.18	1923.07	52.61
2	5	256	1500	0	4.86	3110.49	52.76
3	5	256	1500	0	6.06	1909.46	52.75
4	5	256	1500	0	6.38	1383.41	52.90
5	5	256	1500	0	3.66	2250.79	52.52

2.3.7.2 SAT Solving Based on Clock Glitch Induced Faults

To demonstrate the feasibility of our attack approach, we took as key '0000ffffffffffffffffffffffffffff' and as IV 'ffff2c480000ffffffff0000'. We targeted PRGA round 17. We performed 5 consecutive clock glitch fault injection trials at PRGA rounds 12, 13, 14, 15 and 16 but not at the PRGA round 17. Based on hardware experiments, due to shifting of NLFSR bits, the possible faults are ({127}, 17), ({124}, 17), ({127, 124}, 17), ({126, 125, 124}, 17), ({126, 125, 124, 123}, 17). We passed these faults to the SAT solver as explained earlier. The internal state at PRGA round 17 was successfully recovered in 527.83 seconds with $N = 256$.

2.3.7.3 Probability of Rejecting Bad-Faults

In this chapter any possible fault other than a k-neighborhood fault at a fixed PRGA round (that disturbs only one PRGA round) is considered as a bad-fault. In this chapter we show that arbitrary bad-faults can be rejected in favor of k-neighborhood bit fault. For each column of the following table, we considered a set of 2^{20} experiments (at PRGA round 1).

k	1	2	3	4	5
POS	0.999993	0.999979	0.999963	0.999946	0.999921

Thus bad-faults at PRGA round 1 are rejected with very high probability. Experimental results show that for these cases it is not required to know the actual bit arrangement of the cipher device since faults are identified with high probability.

2.4 Summary

In this chapter, we have presented to the readers a basic understanding of the most classical form of fault attacks on cryptographic implementations - differential fault analysis (DFA). In DFA, an adversary injects a fault under a specific fault model during the operation of the device under test, and obtains the corresponding faulty ciphertext. The correct and faulty ciphertext pairs are then analyzed by the adversary to try and recover one or more sub-parts of the key. Depending on the actual algorithm, the fault location and the fault model may be suitably chosen so as to reduce the key search complexity as much as possible. The key-recovery process usually involves solving a system of equations tracing the fault propagation, and can be expedited using the many efficient SAT-solvers available today. DFA is a very generic attack strategy and is applicable to a whole range of key-based cryptographic primitives, including both public-key cryptosystems, as well symmetric-key cryptosystems such as block and stream ciphers. In the forthcoming section, we demonstrate a slightly different flavor of fault attacks that are either inspired by or assisted by side-channel analysis techniques.

Part II
Side-Channel Inspired and Assisted Fault Analysis Techniques

Chapter 3
Side-Channel Inspired Fault Analysis Techniques

Sikhar Patranabis and Debdeep Mukhopadhyay

Abstract In the previous chapter, we have presented to the reader differential fault attack (DFA) techniques where the adversary analyzes the propagation of an induced fault across the rounds of a block cipher/stream cipher to recover the key. The key recovery process in such attacks usually involves solving a system of equations, tracing the fault propagation characteristics. There exists, in literature, another class of fault attacks that differ from the aforementioned attack strategy in principle. These attacks do not exploit the differential between the fault-free and faulty ciphertexts; rather, they focus on the nature of the fault distribution observed under a variety of fault injection parameters. In many ways, the attack principles are more similar to side-channel analysis in the sense that the observed fault nature serves as a key-dependent leakage, which can then be used to distinguish the correct key guess from the wrong key guesses via a distinguisher. In this chapter, we discuss two such techniques, usually targeting block ciphers - fault sensitivity analysis (FSA) and differential fault intensity analysis (DFIA).

3.1 Fault Sensitivity Analysis of Block Ciphers

Fault sensitivity analysis, abbreviated as FSA, was introduced by Li et al. in [110]. FSA marks a major conceptual departure from the traditional DFA based attack procedures described in the previous chapters. Rather than focusing on the fault propagation characteristics of a given block cipher, FSA exploits the critical fault injection condition when a target device begins to display faulty characteristics. For example, consider a scenario where the adversary gradually increases the fault injection intensity (e.g. voltage spike level or clock glitch frequency), causing the target device to transition from non-faulty behavior to faulty behavior. The critical injection intensity where this crossover takes place is typically data-dependent; consequently, the adversary can use this information to try and retrieve the secret-key by inferring

S. Patranabis (✉) · D. Mukhopadhyay
Secured Embedded Architecture Lab, IIT Kharagpur, India
e-mail: sikhar.patranabis@iitkgp.ac.in

S. Patranabis and D. Mukhopadhyay (eds.), *Fault Tolerant Architectures for Cryptography and Hardware Security*, Computer Architecture and Design Methodologies, https://doi.org/10.1007/978-981-10-1387-4_3

the corresponding sensitive data being processed during the block cipher operation. Quite clearly, FSA is closer in principle to side-channel analysis techniques as compared to traditional DFA techniques.

An important feature of FSA is that the adversary does not explicitly require the knowledge of the faulty ciphertext; the only information required is the critical fault intensity marking the onset of faulty behavior. This makes FSA a potent threat to not only unprotected block cipher implementations, but also countermeasures against DFA [110]. One such example is the Wave Dynamic Differential Logic (WDDL) [175] which, although inherently immune against DFA attacks [164], are vulnerable to FSA. It was, in fact, demonstrated in [110] that a practical ASIC implementation of WDDL-AES is not perfectly secure against the FSA attack based on setup-time violations.

3.1.1 General Attack Principle

The basic attack principle for FSA may be summarized in the following simple steps:

- The adversary, who is in possession of the target device, starts injecting faults of gradually increasing intensity, starting from a fault-free condition to the onset of fault. This is typically manifested in an actual attack by applying voltage/clock glitches, or alternatively EM/laser pulses of gradually increasing magnitude/frequency, as applicable.
- The adversary collects information about the various injection intensities at which the success rate becomes non-zero and also the intensities at which the success rate becomes one, that is, the fault is injected deterministically.
- The adversary then tires to correlate these injection intensities with the internal key-dependent states of the block cipher computation.

3.1.1.1 Data Dependency of Fault Sensitivity

We explain to the reader how fault sensitivity depends on the internal data processed by the block cipher with an example. Consider, for instance, an AND gate with two inputs A and B, and an output C = A$\wedge$B. Let T_{X} denote the timing delay for signal X, and T_{AND} denote the timing delay for the gate itself. Also, let us assume without loss of generality that $T_{\mathrm{A}} < T_{\mathrm{B}}$. Now, it is easy to see that if the value of signal A is zero, we have $T_{\mathrm{C}} = T_{\mathrm{A}} + T_{\mathrm{AND}}$, while if the value of signal A is one, we have $T_{\mathrm{C}} = T_{\mathrm{B}} + T_{\mathrm{AND}}$. In other words, the timing delay for the signal C depends on the value of the signal A. A similar characteristic is shared by an OR gate, where the timing delay for the output signal again depends on the value of the input signal with the shorter delay value. For an XOR gate, however, this is not the case, and the value of the output signal is determined only by the maximum delay of the input signals, irrespective of their value. In other words, *non-linear gates lead to data-dependent timing delay behavior of their outputs.*

Now, consider an adversary injecting faults in a block cipher implementation via clock glitches. The glitch frequency corresponding to the onset of faulty behavior will be determined by the critical timing delay of the circuit, which is essentially the maximum among all data-dependent timing delays in the circuit. This in turn implies that the critical timing delay is also data-dependent. Additionally, it is fixed for a given circuit implemented on a given target device, implying that fault injections at the critical glitch frequency are repeatable with 100% accuracy. Hence, the adversary may inject as many faults as necessary for analyzing the relations between the critical glitch frequency and the secret-key dependent block cipher operations.

Data-dependent critical glitch frequencies are mostly observed in circuits containing non-linear blocks such as S-Boxes, which are crucial to the classical security of nearly all block cipher implementations. Most non-linear blocks comprise of AND gates, which are primarily responsible for data-dependent timing behavior of the overall circuit. In any AND gate, each logic 0 input signal has a probability for decreasing the critical timing delay. Consequently, the more 0s in the input signals, the bigger the possibility that the critical timing delay of the S-Box becomes shorter. This leads to a statistical correlation between the timing delay of the non-linear block and the Hamming weight of the input signals, which are mostly key-dependent in an actual cipher implementation. For example, in [110], the authors presented a case study of the PPRM1 S-Box for AES, proposed by Morioka and Satoh in [124]. The PPRM1 S-Box is vulnerable to FSA attacks due to the presence of an array of non-linear AND gates in its design.

3.1.2 Formal Attack Procedure

The formal procedure for mounting FSA on a block cipher involves two main steps, as described below:

1. **Collection of Sensitivity Information**. The first step for an adversary in an FSA attack is to collect the sensitivity information for the target implementation. Let the corresponding fault injection parameter be denoted as F, where $F = 0$ implies fault-free execution, while larger values of F denote greater injection intensity, such as voltage underfeeding or over-clocking. Let $C = Enc(P, F)$ denote the faulty ciphertext obtained upon injecting a fault with intensity F during encryption of a plaintext P, and let F^c denote the corresponding critical intensity at which the fault begins to occur for the plaintext P. Algorithm 5 summarizes the steps for collecting the fault sensitivity information.
2. **Key-Recovery**. For each faulty ciphertext $C[i]$, the adversary makes several key guesses K_g, predicts a corresponding critical fault injection intensity $F_g^c[i]$, using a prediction function $f(C[i], K_g)$, and correlates the same across all plaintexts with the actual critical fault injection intensity parameters $F^c[i]$ already obtained.

The correlation peak among all possible key guesses is expected to be the same as the correct one. Algorithm 6 summarizes the key-recovery procedure for FSA. Similar to side-channel analysis attacks, the key-hypothesis is made on smaller sub-keys, with the effect of the remaining operations other than the target operation for a specific sub-key treated as noise.

Algorithm 5 FSA: Collection of Fault Sensitivity Information

Input: Cryptographic Algorithm, The number of different plaintexts: N
Output: Ciphertexts: $C[i]$, Critical fault injection intensity: $F^c[i]$ for $i \in \{1, \cdots, N\}$

for $i = 1$ to N **do**
    Generate a random plaintext $P[i]$
    Reset fault injection intensity F to 0
    Initialize $C[i] \leftarrow Enc(P[i], F)$ (correct ciphertext)
    while $Enc(P[i], F) = C[i]$ **do**
        Increase F a little
    end while
    Store $C[i] \leftarrow Enc(P[i], F)$
    Store $F^c[i] \leftarrow F$
end for
return $C[i]$, $F^c[i]$ for $i \in \{1, \cdots, N\}$

Algorithm 6 FSA: Key-Recovery

Input: Bit length of sub-key: t, Faulty Ciphertexts: $C[i]$, Critical Fault Injection intensity: $F^c[i]$ for $i \in \{1, \cdots, N\}$
Output: Correct sub-key K_{correct}

for each $K_g \in \{0, \cdots, 2^t - 1\}$ **do**
    for $i = 1$ to N **do**
        Compute $F_g^c[i] \leftarrow f(C[i], K_g)$
    end for
    Compute the Pearson's correlation coefficient $Corr[K_g] \leftarrow \rho(F^c[i], F_g^c[i])$.
end for
$K_{\text{correct}} \leftarrow K_g$ such that $Corr[K_g]$ is maximum.
return K_{correct}

3.1.3 Advantages of FSA over DFA

We enumerate here some of the advantages offered by FSA over the traditional DFA-based fault injection analysis techniques:

1. FSA does not require the adversary to inject faults under a specific fault model, such as bit-flips or single byte faults. On the other hand, for any DFA, the choice

of fault model plays a crucial role in determining the efficiency of key-recovery. In terms of the number of *successful fault injections* required, DFA outperforms FSA. However, the probability that any fault injection yields a fault under the desired fault model is not necessarily high for all target devices. FSA, on the other hand, does not require the capability of injecting precise faults under a very restricted fault model; it only requires that the adversary correctly identifies the critical fault intensity and correlates the same with sensitive data-dependent operations in the block cipher.

2. FSA attacks do not require the values for faulty ciphertexts. This makes it a potential threat against conventional fault-based attack countermeasures that provide a nonsense output or halt the calculation when a computational fault is detected. The fault sensitivity information is still available to adversaries since they can still identify the critical fault injection intensity.

On the other hand, traditional side-channel countermeasures such as masking, which are usually ineffective against DFA, work well against FSA since they successfully hide all correlation between sensitive-data dependent operations and the critical fault injection intensity. In other words, once all of the intermediate values are masked by uniformly distributed random numbers, fault sensitivity is also randomized and no longer leaks information about the secret-key.

3.2 Differential Fault Intensity Analysis of Block Ciphers

We now introduce a second fault attack technique inspired by side-channel analysis techniques - differential fault intensity analysis [69], abbreviated as DFIA in the rest of the discussion. DFIA is based on the fact that practical fault injections on real-life target devices always result in a non-uniform/biased fault distribution. The non-uniformity can be observed across a wide variety of fault injection intensities. For example, when injecting faults on AES-128 at low fault intensities, small perturbations such as single/double bit flips are more likely occur in the target bytes, as opposed to uniformly random faults of larger Hamming weight. DFIA exploits this inherent bias in fault distribution, observed across multiple fault injections, by correlating the same with a hypothesis for the secret-key. The adversary performs several fault injections under a given fault intensity, obtains the corresponding faulty ciphertexts, and derives an intermediate, key-dependent state variable under a given key hypothesis. A distinguisher is then applied to identify the key hypothesis for which, the state variable most accurately reflects the non-uniform fault distribution pattern. In particular, due to the presence of a non-linear layer in most block ciphers, wrong key hypotheses would always lead to inference of states with uniform high-Hamming weight fault distributions. This is very similar in principle to differential power analysis (DPA), where the adversary derives intermediate state variables under different key hypotheses, and correlates the corresponding simulated power leakage with the observed leakage for the target device.

3.2.1 The Fault Model for DFIA

As already mentioned, DFIA depends heavily on the bias of injected faults in practice. The original proponents of DFIA [69] assumed that a single byte fault was injected in the cipher state, and that the distribution of byte faults was non-uniform, implying that not all byte faults occur with equal probability. It was also assumed that the number of faulty bits is dependent on the fault intensity, which is a controllable parameter from an adversary's point of view. Finally, as is the case with most fault attacks the adversary is assumed to have full knowledge of the fault location, for example, the target byte of the cipher state. Since an adversary can always characterize the required timing of fault injection from the side-channel leakage of an implementation, such as assumption is reasonable to make.

3.2.2 The General Attack Methodology

We now present a generic description of the DFIA methodology. We assume that the adversary is aware of the cryptographic algorithm implemented in the device under test, and that he is able to inject a fault into a single byte of an intermediate state S of the algorithm, such that the fault propagates to a key-dependent observable output. Finally it is also assumed that the adversary can control the intensity of the injected fault up to a granularity of 1 bit. The main steps of the DFIA attack are summarized below:

- **Step-1: Fault Injection**. The adversary performs Q executions of the encryption algorithm on the same plaintext P and the same secret key K. In each execution instance $q \in [1, Q]$, he introduces a single byte fault f_q into the same intermediate state S of the encryption algorithm, resulting in a faulty intermediate state S'_q. The fault propagates to the output and the adversary collects the corresponding faulty ciphertext C'_q. The adversary can vary the intensity of the fault injection across these Q executions - it is not necessary for the adversary to obtain the correct ciphertext C in the process.
- **Step-2: Key Recovery**. Note that from the knowledge of the faulty ciphertext C'_q, the adversary can infer a faulty secret state $S'_{k,q}$ by hypothesizing on the part of the secret key k, that was originally combined with the faulty state S'_q to obtain C'_q. For each possible hypothesis of k, the adversary will infer a different $S'_{k,q}$. For the correct key guess, $S'_{k,q}$ will match the original faulty intermediate state S'_q, while for a wrong key guess, $S'_{k,q}$ will be uniformly random. Now, if the distribution of the faults $f_1, f_2, \ldots, f_Q$ is biased towards a specific value, then for a correct key guess k, the distribution of the inferred intermediate states $S'_{k,1}, \ldots, S'_{k,Q}$ will also be biased. On the other hand, for a wrong key guess, the corresponding distribution will be uniformly random, with each inferred state independently and identically distributed. Hence, after formulating the inferred intermediate state variables $S'_{k,1}, \ldots, S'_{k,Q}$ for each key hypothesis k, the adversary can apply an

appropriate statistical distinguisher to differentiate the biased distribution from the remaining uniformly random ones. One such distinguisher is the Hamming Distance HD between the pairwise intermediate states. If the fault distribution is indeed biased, with a very high fraction of the injected faults concentrated in a particular Hamming Weight zone, then the expression $\sigma_k = \sum_{q_1,q_2=1}^{Q} \text{HD}(S'_{k,q_1}, S'_{k,q_2})$ is expected to be minimum for the correct key hypothesis k.

The above methodology can now be used to infer a divide-and-conquer strategy, wherein the adversary can inject faults in different bytes of the intermediate state S, and use the resulting fault distributions across multiple such fault injection instances to recover different parts of the entire secret key K. Note the inherent similarity of DFIA with DPA - in both forms of attack, the adversary assumes a certain model of leakage (e.g. Hamming Weight/Hamming Distance model in case of DPA and an appropriate biased fault model in DFIA), hypothesize on a part of the key that links this leakage to the observable output to infer an intermediate state, and then uses an appropriate metric/distinguisher to rule out the wrong key guesses. One can even say the following: *DFIA is essentially an adaptation of DPA in the context of fault analysis*. Algorithm 7 summarizes the generic steps for DFIA on any target cipher implementation.

Algorithm 7 DFIA-Based Attack: A Generic Overview

```
Input: Cryptographic Algorithm, Fault Injection Methodology InjectFault
Output: Correct Key Guess for K
InjectFault (Target Round, Target Location, Q)
Obtain Faulty Ciphertexts C'_1, C'_2, ..., C'_Q
for each fault f_q where q ∈ [1, Q] do
    for each key hypothesis k do
        Let C'_q be the faulty ciphertext
        Infer intermediate state variable S'_{k,q}
    end for
end for
for each key hypothesis k do
    Compute σ_k = Σ_{q1,q2=1}^{Q} HD(S'_{k,q1}, S'_{k,q2})
end for
Set K = argmin_k {σ_k}
```

3.2.3 *Advantages of DFIA*

As compared to the fault attacks introduced previously, DFIA seems to offer two major advantages in terms of ease of injection and exploitability requirements, which are enumerated below:

1. Unlike DFA, DFIA does not require an extensive analysis of the fault propagation characteristics for the target cipher. Recall that in DFA, the adversary essentially compares the outputs of faulty and fault-free ciphertexts corresponding to the same plaintext under a known fault model to try and infer the secret key. In DFIA, on the other hand, it is enough for the adversary to have access to faulty ciphertexts under varying fault intensities. In particular, *DFIA does not require access to correct ciphertexts.*
2. Unlike fault sensitivity analysis (FSA), DFIA does not require identification of the exact critical fault injection intensity. DFIA essentially tries to infer the secret state of the target cipher purely based on the knowledge of fault distributions and algorithmic responses, and does not depend on the fault sensitivity of the device.

3.2.4 Case Study: DFIA on AES-128

We present a case-study here on DFIA of AES-128. Recall that AES-128 has 10 rounds preceded by a key-whitening step, such that each round from 1 to 9 comprises of four operations - SubByte, ShiftRow, MixColumn and AddRoundKey, while the last round has the same set of operations with the exception of MixColumn. Now, suppose an adversary wishes to mount a DFIA on AES-128. He may proceed as follows:

1. The target state S for the adversary for fault injection is the cipher state at the beginning of round 10 (the last round) of AES.
2. The adversary performs Q fault injections $f_1, \ldots, f_Q$, and collects the faulty ciphertexts $C'_1, \ldots, C'_Q$ corresponding to the same plaintext P.
3. Consider the qth fault injection instance. Since the adversary injects the fault f_q after the 9th round MixColumn operation, and the last round is entirely devoid of MixColumn, the fault stays localized within a single byte of the ciphertext.
4. Consequently, the adversary makes a key hypothesis $k \in [0, 255]$ on the corresponding byte of the last round key of AES and infers the faulty state byte $S'_{k,q}$ as:
$$S'_{k,q} = \mathsf{InvShiftRow}\left(\mathsf{InvSubByte}\left(C'_q \oplus k\right)\right)$$
5. Finally, the correct key byte K is computed as $K = \text{argmin}_k\{\sum_{q_1,q_2=1}^{Q} \text{HD}(S'_{k,q_1}, S'_{k,q_2})\}$ where $k \in [0, 255]$.

Thus the complexity of recovering a single byte of the last round key of AES-128 is 2^8. Applying the same fault injection and key recovery procedure to all bytes, the adversary is thus expected to recover the entire key of AES-128 with an attack complexity of 2^12.

We present some simulation results for DFIA on AES-128 in Fig. 3.1. The figure depicts the number of fault injections required versus the number of key bits recovered for various combination of fault models. In particular, we depict DFIA using only

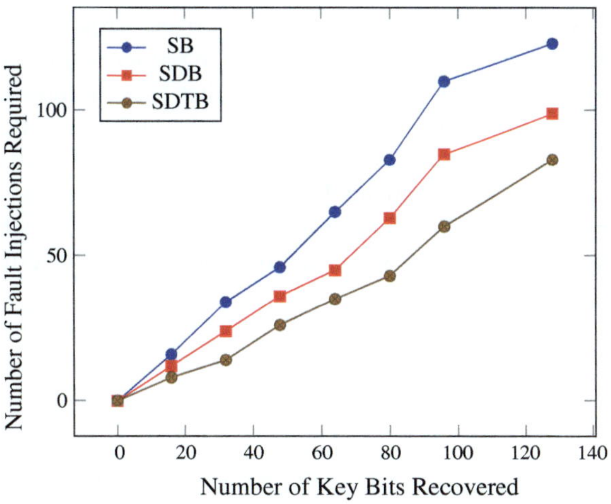

Fig. 3.1 Simulation Result: DFIA on AES-128: Fault Injections versus Key Bits Recovered

single bit faults as SB, DFIA using a combination of single and double bit faults as SDB, and DFIA using a combination of single, double and triple bit faults as SDTB, respectively.

3.3 Chapter Summary

In this chapter, we have presented to the readers two slightly unconventional fault attack techniques that are more similar in philosophy to side-channel attack techniques such as DPA. The first of these- fault sensitivity analysis (FSA) essentially heralded this shift of paradigm from traditional DFA to more implementation-specific side-channel-like attacks. In FSA, the adversary tries to correlate the critical fault intensity, marking the onset of faults in a given block cipher implementation, with the key-dependent internal state of the cipher algorithm. This is, in principle, very similar to DPA where the power consumption of the device under test is correlated with the key-dependent cipher state. The correct key is then identified as the peak of all correlation values for different key hypotheses. FSA is probably the first reported fault attack technique that requires neither the faulty ciphertext values, nor specific fault models, to recover the key. It is therefore a potent threat to detection based countermeasure techniques against DFA, that often try to either suppress/randomize the faulty ciphertexts. On the flip side, any implementation protected against side-channel attacks also resists FSA, but not necessarily DFA.

In many ways, differential fault intensity analysis (DFIA) is a mixture FSA and DFA. Instead of fault sensitivity, it focuses more on the exact distribution of the faulty ciphertexts under a given fault model, and exploits the fault bias to try and recover the key. DFIA requires only faulty ciphertexts, and uses an appropriate distinguisher to try and recover the secret key. While the use of a specific fault model for the attack is similar to DFA, the final key recovery process is similar to FSA (and hence, in principle, to side-channel attacks) in the sense that the adversary makes several key hypotheses, and chooses the one that best captures the biased nature of the faulty cipher state from the knowledge of the faulty ciphertexts.

In conclusion, both FSA and DFIA represent a different class of fault attacks, that are closer to side-channel analysis in spirit rather than traditional fault analysis. They require more fault injections as compared to DFA attacks described earlier, but the key-recovery process is usually simpler and more efficient.

Chapter 4
Side-Channel Assisted Fault Analysis

Sikhar Patranabis, Jakub Breier, Debdeep Mukhopadhyay and Shivam Bhasin

Abstract In the preceding chapters, we have examined fault attacks as a form of active side-channel analysis, wherein the adversary injects a fault, and traces its propagation characteristics to recover the key. Certain classes of fault attacks such as FSA and DFIA have attack principles which are similar to that of side-channel attacks. In this chapter, we introduce to the reader a novel form of fault attack referred to as *side-channel assisted fault attack*. Such attacks demonstrate the combined power of side-channel and fault attacks working in tandem to recover the secret key from a given block cipher implementation. The attack is demonstrated on the current worldwide standard for lightweight block ciphers—PRESENT, which uses a substitution-permutation network (SPN) spread across 31 rounds, and affords 80/128 bit-security for a plaintext block of size 64.

4.1 Introduction

The design of PRESENT [34] has inspired a whole new family of block ciphers that use bit-permutations as opposed to maximum distance separable (MDS) layers for achieving the necessary diffusion characteristics. Bit-permutations afford zero-cost implementations in hardware as they can be realized solely via wiring. Several new proposals for lightweight block ciphers such as Rectangle [182] and GIFT [16] have opted for bit-permutations for efficiency in hardware. In this chapter, we propose a novel side-channel assisted fault analysis of PRESENT and PRESENT-like ciphers that exposes an interesting vulnerability of bit-permutations. In particular, the attack is based on the principle that for bit-permutation based ciphers, the knowledge of the

S. Patranabis (✉) · D. Mukhopadhyay
Secured Embedded Architecture Lab, IIT Kharagpur, Kharagpur, India
e-mail: sikhar.patranabis@iitkgp.ac.in

J. Breier · S. Bhasin
Physical Analysis and Cryptographic Engineering, Temasek Laboratories, Nanyang Technological University, Singapore, Singapore

S. Patranabis and D. Mukhopadhyay (eds.), *Fault Tolerant Architectures for Cryptography and Hardware Security*, Computer Architecture and Design Methodologies, https://doi.org/10.1007/978-981-10-1387-4_4

output fault mask in an earlier round could be exploited to significantly reduce the entropy of the input fault mask at a later round.

The chapter introduces to the reader a practically realizable combination of side-channel analysis (SCA) and differential fault analysis (DFA) on PRESENT. The attack uses a relaxed fault model corresponding to a given target round, and assumes that the adversary uses side-channel leakages to precisely determine the resulting fault mask. Prior attacks on PRESENT have mostly been limited to either SCA or FA, but have not exploited the combined potential of both. We present a theoretical analysis to establish that the attack is capable of recovering multiple key nibbles in parallel under the same fault injection instance, implying that the attack is not only feasible but also efficient.

We further corroborate the theoretical analysis with a real-world demonstration of the combined attack on an ATmega328P microcontroller-based implementation of PRESENT-80 using a laser-driven fault injection setup. Our experiments demonstrate that the attack recovers 64 bits of the last round key of PRESENT using only 4 fault injections in the best case, and 7–8 fault injections in the average case. This makes the attack one of the most efficient to be proposed on PRESENT and PRESENT-like block ciphers.

The combined attack is, in principle, aided by the diffusion characteristics of PRESENT. This sheds light on an hitherto unexploited weakness of bit-permutations: the knowledge of the output mask of a suitable injected fault in an earlier round can be used to significantly reduce the entropy of the output fault mask in a subsequent round, even if the two rounds are executed as many as three clock cycles apart. The same attack would not, however, apply to an MDS-layer based block cipher such as AES, where such fault propagation characteristics are absent. To the best of our knowledge, this is the first attack to demonstrate this inherent vulnerability of PRESENT and PRESENT-like block ciphers using bit-permutations for implementation efficiency. The reader is encouraged to dig deeper into the possibility of exploring similar vulnerabilities that may be inherent in bit-permutation based block cipher designs.

4.2 Chapter Preliminaries

4.2.1 Overview of the PRESENT Block Cipher

PRESENT is based on a substitution-permutation network (SPN). It consists of 31 rounds, block length is 64 bits and it supports keys with lengths of 80 and 128 bits. In this chapter, we focus on the 80 bit key length version, which we denote as PRESENT-80, however the attack applies on 128-bit as well. Each round consists of three operation layers: an XOR-layer with the round key, a substitution layer using 16 identical 4×4 S-Box (Table 4.1) and a bit-permutation layer (Fig. 4.1). At the end of round 31, a post-whitening XOR with the round key is performed, so there

Table 4.1 The PRESENT S-Box

x	0	1	2	3	4	5	6	7	8	9	A	B	C	D	E	F
$S[x]$	C	5	6	B	9	0	A	D	3	E	F	8	4	7	1	2

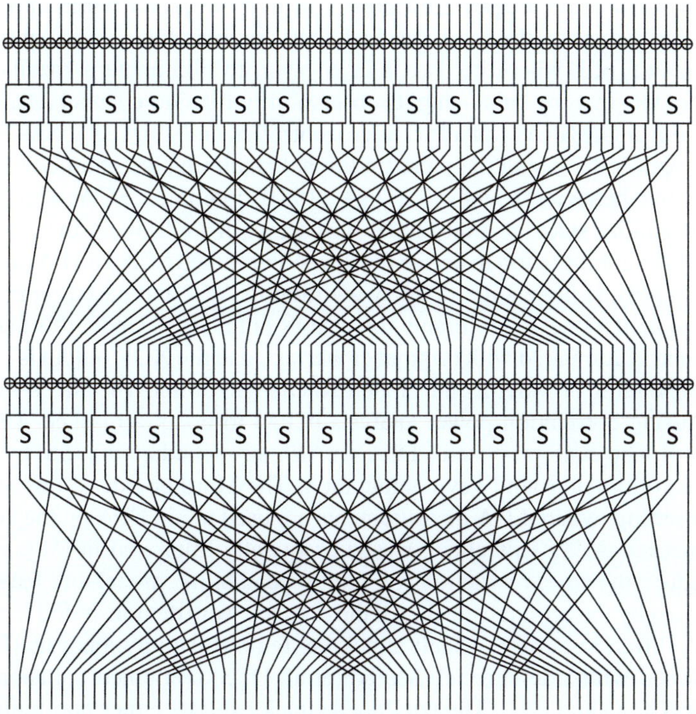

Fig. 4.1 Structure of two rounds of PRESENT

Table 4.2 Summary of combined SCA and FA on SKC

Reference	Setting	Target	Combination
Robisson et al [155]	Simulated	Non-masked AES	DPA + Stuck-At Fault
Clavier et al. [48]	Simulated	Masked AES	CPA + Instruction-skip Fault
Roche et al. [156]	Simulated	Key schedule of masked AES	CPA + Stuck-At or Byte Fault
Dassance et al. [58]	Simulated	Key schedule of masked AES	CPA + Byte Fault
Moradi et al. [123]	Practical	Unprotected + Protected AES	CCA + FSA
Li et al. [108]	Practical	Unprotected AES	DPA + FSA

are 32 generated round keys in total. The key schedule for PRESENT-80 comprises of a rotation, S-Box look-up and round counter addition, thus making it invertible (Table 4.2).

4.2.2 Associated Literature

4.2.2.1 Combined SCA and FA on Symmetric-Key Cryptosystems

While both SCA and FA are powerful attacks on their own, some works have explored the combined power of the two. The first such combined attack was proposed by Skorobogatov [166], where he used a laser beam to illuminate a specific area on the chip surface, thereby enhancing its observability for side-channel attacks. It was demonstrated on the SRAM of a PIC microcontroller, and allowed access to individual bits as opposed to using the more traditional Hamming weight model.

Differential behavioral analysis (DBA [155]) combined SCA with safe-error attacks. Assuming stack-at fault model it observes if fault alters the side-channel behavior of the computation to derive the key. A combined SCA and FA on AES was proposed in [48]. It targets the first key addition in AES and based on instruction-skip/change fault model to preferably force XOR output to 0. Under this fault model, the ciphertext is compared with original ciphertext, and the XOR output is inferred to be 1 or 0 depending on whether the ciphertext changes or not. The attack was further enhanced using correlation power analysis (CPA) to break a masked AES implementation. Roche et al. proposed a DFA on AES key schedule in [156] by injecting faults in pen-ultimate round key computation. They further extend this attack to a combined setting, where SCA measurements are used to aid DFA on the key schedule of a masked AES. This attack was subsequently improved in [58], where the authors reduce the strict restrictions on fault repeatability, model and location, that were imposed by the original attack. All these attack were demonstrated in simulated settings. A different family of fault attack i.e. fault sensitivity analysis (FSA) was also combined with side-channel. Moradi et al. [123] combined collision correlation attack (CCA) and FSA. The combined attack exploits either non-uniform fault distribution or data-dependent timing of faults, and was successfully demonstrated on several unprotected and protected AES cores on SASEBO LSI chips. In another work [108], the authors use FSA to develop a leakage model which is then used to launch a power based key recovery attack. Both these attacks were demonstrated with real measurements. To the best our knowledge, no previous work demonstrates a practical attack combining DFA with side-channel.

4.2.2.2 DFA on PRESENT

The first DFA on PRESENT, published by G. Wang and S. Wang [176] required 64 pairs of correct and faulty ciphertext on average, with a computational complexity of 2^{29}. Later, Zhao et al. [183] utilized a fault-propagation pattern-based DFA, targeting PRESENT and PRINT. The attack on PRESENT-80/128 required 8/16 ciphertext pairs on average, reducing the key search space to $2^{14.7}/2^{21.1}$ in average. Bagheri et al. [13] showed attacks utilizing single bit-flip and single nibble fault models, while the second attack revealed the key with 18 ciphertext pairs on average. Breier and He [40] proposed a multiple fault attack, targeting four nibbles at once, being able to recover the secret key in 2 encryptions. DeSantis et al. [59] presented a ciphertext-only attack, which requires only two ciphertext pairs in the best case. Ghalaty et al. [68] attacked PRESENT and LED ciphers with Differential Fault Intensity Analysis (DFIA), showing that both ciphers can be broken with a practically feasible number of fault injections.

4.3 The Combined SCA and DFA of PRESENT

In this section, we present to the reader a combined SCA and DFA of PRESENT. For clarity of presentation, we consider a slightly modified version of the cipher, where we ignore the bit-permutation operation of round 31. It is easy to see that upon recovering the last round key corresponding to this version of PRESENT, one can easily re-apply the inverse permutation layer to obtain the original last round key. We now present a detailed description of the fault attack.

4.3.1 Properties of the PRESENT Block Cipher

We begin with a description of the properties of PRESENT that are exploited in the attack. We would like to point out that these properties are due to the bit-permutation layer of PRESENT, and are also observed in other PRESENT-like block ciphers that use bit-permutations as opposed to MDS layers for diffusion.

- The input of an S-Box in round r comprises output bits from four different S-Boxes in round $r - 1$.
- The output of an S-Box in round r is distributed across the inputs of four different S-Boxes in round $r + 1$.
- The distribution of inputs and outputs across consecutive rounds occurs among four distinct groups of S-Boxes:
  - The output of S-Boxes 0–3 in round r constitute the inputs for the S-Boxes $0, 4, 8$ and 12 in round $r + 1$.
  - The output of S-Boxes 4–7 in round r constitute the inputs for the S-Boxes $1, 5, 9$ and 13 in round $r + 1$.
  - The output of S-Boxes 8–11 in round r constitute the inputs for the S-Boxes $2, 6, 10$ and 14 in round $r + 1$.
  - The output of S-Boxes 12–15 in round r constitute the inputs for the S-Boxes $3, 7, 11$ and 15 in round $r + 1$.

In general, the output of the S-Box group $[4n, 4n + 3]$ in round r entirely constitutes the input for the S-Box group $\{n, n + 4, n + 8, n + 12\}$ in round $r + 1$, where $n \in \{0, 1, 2, 3\}$. More precisely, for $n, d, l \in \{0, 1, 2, 3\}$, the lth bit in the output of S-Box

$4n + d$ in any round is precisely the dth bit in the input of S-Box $n + 4l$ in the next round, albeit after XOR-ing with the corresponding round key bit. As stated before, visualization of this propagation is depicted in Fig. 4.1.

4.3.2 Fault Model and Fault Location

The attack assumes random nibble fault model. The fault is injected at the input of round 28 during a PRESENT encryption operation. Nibble faults have been demonstrated to be practically achievable using traditional fault injection techniques such as clock and voltage glitches [2, 19, 20] as well as more advanced injections methods such as EM pulse or laser pulse injection [61, 125, 171]. We use the term *output fault mask* to denote the differential Δ_{out} of the correct and faulty nibble at output of S-Box operation in round 28. We would like to point out that recent attacks on PRESENT, such as that presented by Breier et al. in [41], assume some specific instances of nibble faults that result in a desired output fault mask. The attack, on the other hand, assumes a random nibble fault, without any specific requirements on the nature of the corresponding output fault mask. This makes our fault model more relevant in the context of real-world fault attacks.

4.3.3 The Role of Side-Channel Analysis in Our Attack

The role of the side-channel leakage in our analysis is to deterministically obtain the output fault mask Δ_{out} corresponding to round 28. Note that since the attack assumes a random fault in a single nibble, the output fault mask Δ_{out} is not known. Instead, we use side-channel analysis to determine Δ_{out}. The basic principle is as follows: assuming the fault is injected in a particular nibble during round r (round 28 in the attack on PRESENT), we collect the leakage traces corresponding to the fault-free and faulty computations in round $r + 1$ (round 29 in the attack). By computing the difference in side-channel measurements of correct and faulty execution, the bit-permutation reveals makes it possible to determine the exact value of Δ_{out}. We practically demonstrate the recovery of Δ_{out}, when injecting faults using a laser fault injection a 8-bit micro-controller platform (see Sect. 4.4 for more details on the experimental set-up). The injected fault was injected during the S-Box lookup corresponding to the target nibble in round 28 of PRESENT encryption. The fault injection timings were profiled with respect to each nibble operation, allowing a 100% repeatability in corrupting a target nibble. This was followed by a differential analysis between the leakage traces corresponding to the fault-free and faulty computations to retrieve the output fault mask. Detailed experimental results corresponding to the SCA are presented in Sect. 4.4.

4.3.4 The Fault Propagation Characteristics

We now present the fault propagation characteristics corresponding to the attack in details. As already mentioned, the attack targets a single nibble in round 28 of PRESENT. We begin by stating the following theorems.

Theorem 4.2 *Suppose the Hamming Weight of the output fault mask of the target nibble in round* 28 *of PRESENT is x, where $x \in \{0, 1, \cdots, 4\}$. Then, the Hamming Weight of the input fault mask of any nibble in round* 31 *is at most x.*

Theorem 4.3 *Suppose the Hamming Weight of the output fault mask of the target nibble in round* 28 *of PRESENT is x, where $x \in \{0, 1, \cdots, 4\}$. Then, the input fault mask of any nibble in round* 31 *takes at most 2^x values.*

We first present two specific examples to support the above theorems, and then prove them for generalized instances.

4.3.4.1 Example-1

Suppose the output fault mask of the target nibble in round 28 of PRESENT is 0001. This implies that in round 29, nibble 0 has an input fault mask of 0001. Unfortunately, the corresponding output fault mask is non-deterministic: all one can infer is that each of the nibbles 0, 4, 8 and 12 in round 30 have an input fault mask of 0000 (implying no fault propagation) or 0001 (implying fault propagation). Once again, the output fault masks in round 30 are non-deterministic; however, one can easily make the following observations:

- If the input fault mask for nibble 0 in round 30 is 0001, then the input fault mask for nibbles 0, 4, 8 and 12 in round 31 are either 0000 or 0001. On the other hand, if the input fault mask for nibble 0 in round 30 is 0000, then the input fault mask for nibbles 0, 4, 8 and 12 in round 31 is definitely 0000.
- If the input fault mask for nibble 4 in round 30 is 0001, then the input fault mask for nibbles 1, 5, 9 and 13 in round 31 are either 0000 or 0001. The case of input fault mask 0000 follows analogously.
- If the input fault mask for nibble 8 in round 30 is 0001, then the input fault mask for nibbles 2, 6, 10 and 14 in round 31 are either 0000 or 0001. The case of input fault mask 0000 follows analogously.
- Finally, if the input fault mask for nibble 12 in round 30 is 0001, then the input fault mask for nibbles 3, 7, 11 and 15 in round 31 are either 0000 or 0001. The case of input fault mask 0000 follows analogously.

Thus, for each of the nibbles in round 31, the input fault mask is either 0000 or 0001, and has Hamming weight at most 1. Figure 4.2 illustrates the fault propagation characteristics for Example-1.

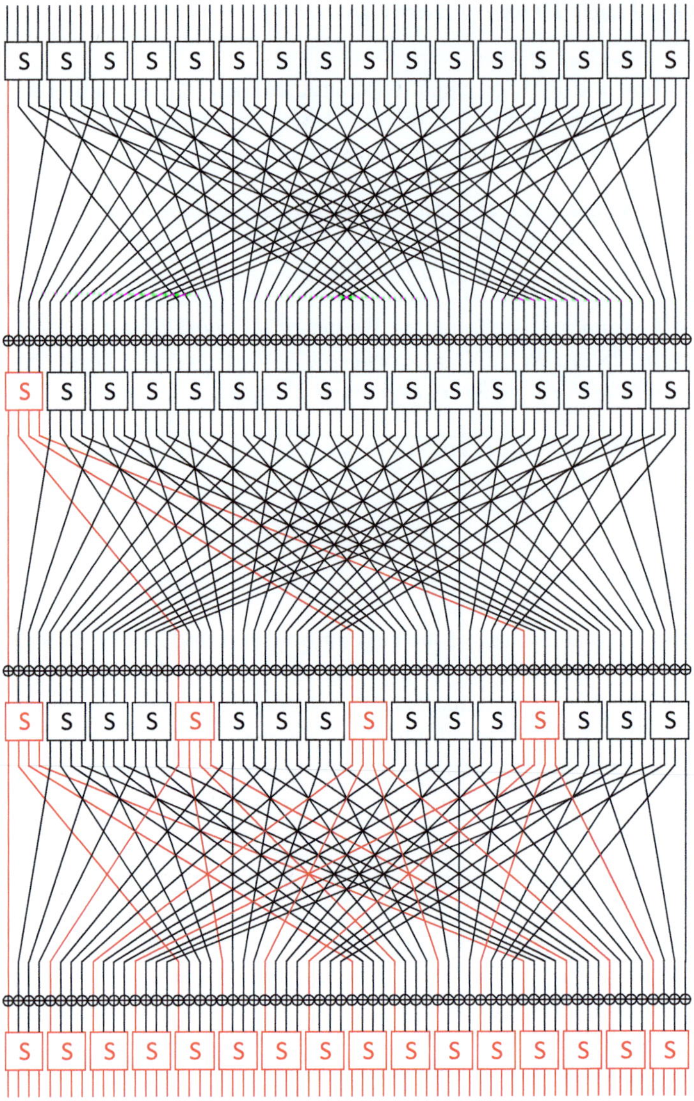

Fig. 4.2 Fault propagation for the output fault mask 0001 (Example-1)

4.3.4.2 Example-2

Suppose the output fault mask of the target nibble in round 28 of PRESENT is 0011. This implies that in round 29, nibble 0 and nibble 4 has an input fault mask of 0001. Once again, once can infer the following:

- Each of the nibbles 0, 4, 8 and 12 in round 30 have an input fault mask of 0000 (implying no fault propagation) or 0001 (implying fault propagation).
- Similarly, each of the nibbles 1, 5, 9 and 13 in round 30 have an input fault mask of 0000 (implying no fault propagation) or 0001 (implying fault propagation).

The analysis in case of round 31 now becomes more complicated. So we focus on the input fault mask for a specific set of nibbles –0, 4, 8 and 12 respectively:

- If the input fault mask for both nibble 0 and nibble 1 in round 30 is 0000, then the input fault mask for nibbles 0, 4, 8 and 12 in round 31 is definitely 0000.
- If the input fault mask for nibble 0 and nibble 1 in round 30 are 0000 and 0001, respectively (equivalently 0001 and 0000, respectively), the input fault mask for nibbles 0, 4, 8 and 12 in round 31 is 0001 (equivalently 0010).
- Finally, if the input fault mask for both nibble 0 and nibble 1 in round 30 is 0001, then the input fault mask for nibbles 0, 4, 8 and 12 in round 31 is definitely 0011.

The input fault mask for the other sets of nibbles may be similarly characterized. Thus, once again, the input fault mask for each of the nibbles in round 30 has Hamming Weight at most 2, and takes at most $2^2 = 4$ values. Figure 4.3 illustrates the fault propagation characteristics for Example-2.

4.3.4.3 The Generalized Proof

The above examples provide an intuitive explanation for Theorems 4.2 and 4.3. We now present a generalized and formal proof for the same. We present the fault mask characteristics in each of rounds 28, 29, 30 and 31 separately.

- **Round 28**: Suppose the adversary injects a fault in nibble $4n + d$, where $n, d \in \{0, 1, 2, 3\}$, and suppose the output fault mask has Hamming weight $x \in \{0, 1, \cdots, 4\}$. In particular, let $l_1, \cdots, l_x \in \{0, 1, 2, 3\}$ be the bits in the output fault mask that are set to 1.
- **Round 29**: Consider the effects of the bit $l_1, \cdots, l_x$ in the output fault mask corresponding to nibble $4n + d$ in round 28. As per the generic properties of the diffusion layer of PRESENT discussed in Sect. 4.3.1, these faulty bits will propagate to the nibbles $n + 4l_1, \cdots, n + 4l_x$ respectively, creating an input fault mask of Hamming Weight 1.
- **Round 30**: We now focus on a specific faulty nibble, say nibble $n + 4l_1$, in round 29. Once again, as per the generic properties of the diffusion layer of PRESENT discussed in Sect. 4.3.1, the output of this faulty nibble will potentially propagate to the nth input bit of the nibbles $l_1, l_1 + 4, l_1 + 8$ and $l_1 + 12$ in round 30. Similarly, the output of the faulty nibble $n + l_x$ in round 29 will potentially affect the nibbles $l_x, l_x + 4, l_x + 8$ and $l_x + 12$. *It is important to observe that for each of* $l_1, \cdots, l_x$, *the set of potentially faulty nibbles in round 30 is disjoint.*
- **Round 31**: Finally, we examine the fault propagation to the input of round 31. Consider the faulty quartet of nibbles $l_1, l_1 + 4, l_1 + 8$ and $l_1 + 12$ in round 30. The nibble l_1 will potentially spread the fault to the l_1^{th} input bit of the nibbles 0, 4, 8

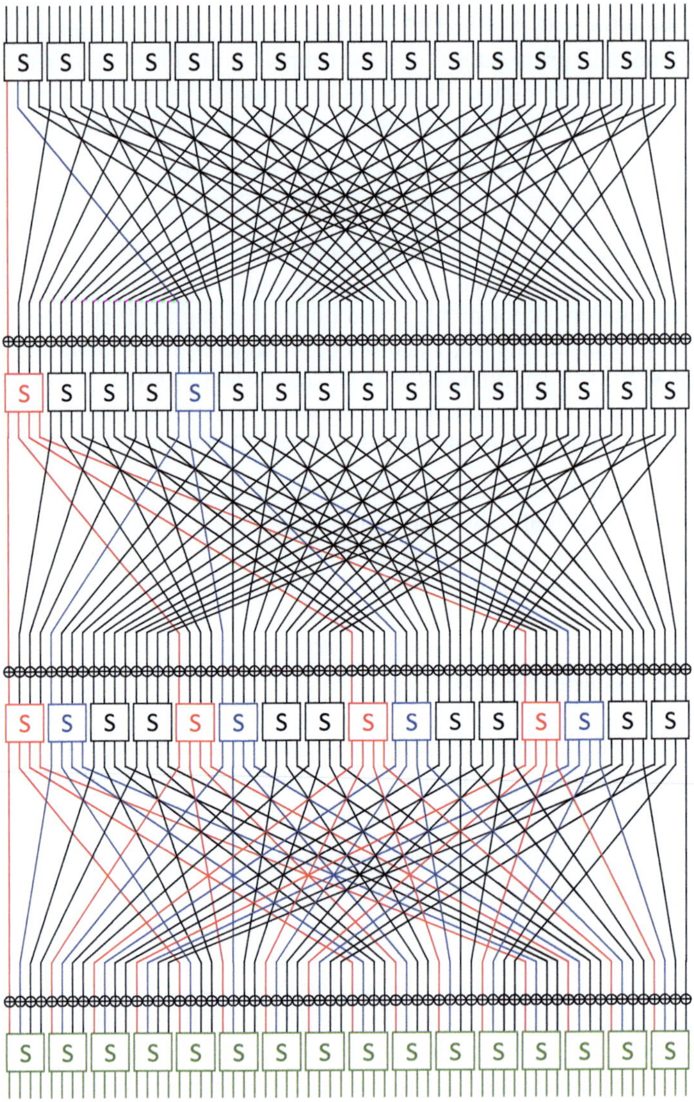

Fig. 4.3 Fault propagation for the output fault mask 0011 (Example-2)

and 12 in round 31. Similarly, the nibble $l_1 + 4$ will potentially spread the fault to the l_1^{th} input bit of the nibbles 1, 5, 9 and 13 in round 31. In general, the faulty nibble $l_{k_1} + 4k_2$, where $k_1 \in \{1, \cdots, x\}$ and $k_2 \in \{0, 1, 2, 3\}$, will potentially spread the fault to the $l_{k_1}^{\text{th}}$ input bit of the nibbles k_2, $k_2 + 4$, $k_2 + 8$ and $k_2 + 12$ in round 31. This, in turn, implies each nibble in round 31 can receive a faulty input bit from at most x faulty nibbles in round 30.

Thus, each nibble in round 31 has an input fault mask of Hamming Weight at most x. Additionally, since exactly x bits of each input fault mask are 1, and the faulty bits are determined by the values of $l_1, \cdots, l_x$, each input fault mask in round 31 can take 2^x values. This completes the proof of Theorems 4.2 and 4.3.

4.3.5 The Key Recovery Process

The fault propagation characteristics described above can now be used to recover multiple key nibbles in parallel. Recall that in the modified version of PRESENT which we consider for the attack, we ignore the bit-permutation layer in round 31, implying that for any nibble with a non-zero input fault mask, the output is directly XOR-ed with the last round key nibble and output as the ciphertext. Thus, given a correct ciphertext nibble C and a faulty ciphertext nibble C', corresponding to a non-zero input fault mask β, we have the following differential relation involving the corresponding the final round key nibble K:

$$S^{-1}[C \oplus K] \oplus S^{-1}\left[C' \oplus K\right] = \beta$$

where S^{-1} denotes the inverse S-Box operation. From the differential uniformity of the PRESENT S-Box, the expected number of values of K satisfying the above equation is one. Now, assuming a non-zero output fault mask with Hamming weight x for the target nibble in round 28, there are $2^x - 1$ possible values of the input fault mask β in round 31 (this is proven using Theorems 4.2 and 4.3), which in turn gives rise to $2^x - 1$ possible differential relations as described above. Hence, any value of x in the set $\{1, 2, 3\}$ reduces the entropy of the key nibble K by a factor of approximately 2^{4-x} on an average, and is hence expected to allow recovering K uniquely after $4/(4 - x)$ fault injections. For $x = 1, 2$ and 3, the expected number of fault injections are thus 1.33, 2 and 4 respectively. The overall attack flow is summarized in Fig. 4.4.

It is also worth observing that we do not require separate fault injection instances for recovering each key nibble. On the contrary, each fault injection instance in round 28 is expected to yield multiple faulty nibbles in round 31, and each of these nibbles may be analyzed independently and in parallel for key recovery. This inherent parallelism makes the attack very efficient and reduces the overall number of fault injections necessary to recover 64 bits of the last round key. The attack efficiency is also supported by experimental results in Sect. 4.4. Like any other side-channel or fault attack on PRESENT, the remaining 16 bits of the 80 bit key may be brute-forced even in this attack. An alternative strategy could be to retrieve 64 bits of both the last round key as well as the penultimate round key by repeating the attack twice; once in round 28 and then in round 27. The combination of the two round keys can then be used to recover the entire last round key of PRESENT.

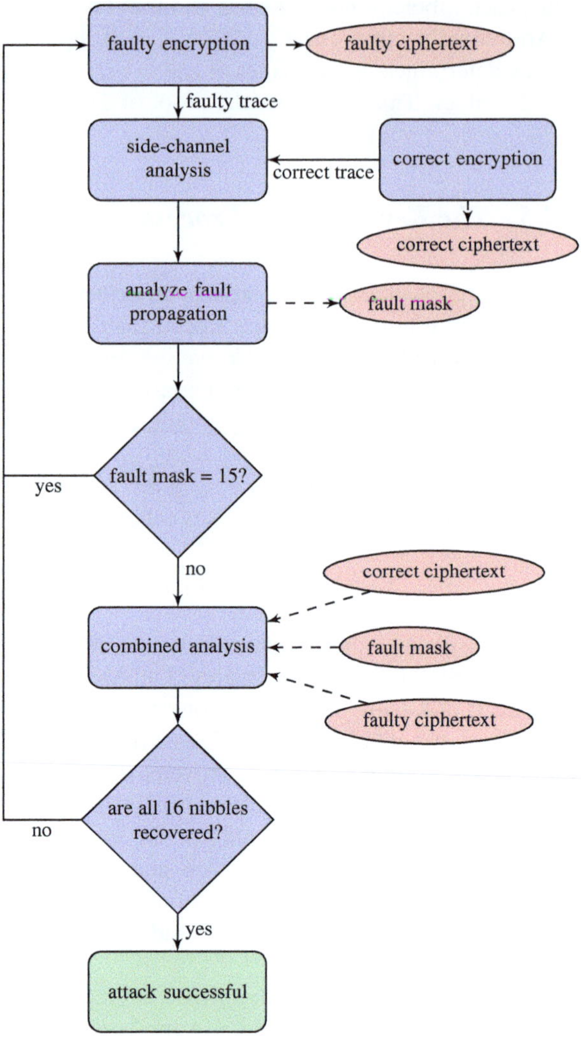

Fig. 4.4 Summary of the Combined SCA+DFA on PRESENT

4.4 Experimental Results

4.4.1 The Combined SCA+FA Setup

The setup for our combined SCA and DFA-based attack is depicted in Fig. 4.5. The core of the setup consists of near-infrared diode pulse laser (1064 nm wavelength) with the maximum output power of 20 W. This power is further reduced to ≈8 W with usage of 20× objective lens that scales the effective spot size to 15 × 3.5 μm.

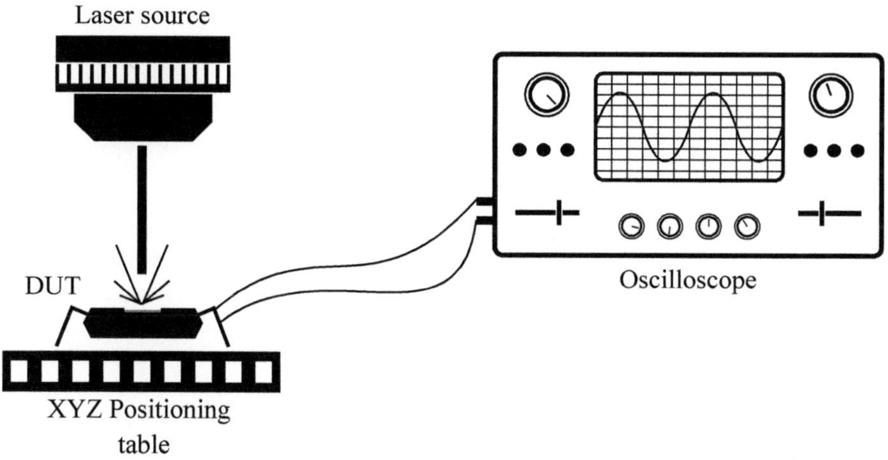

Fig. 4.5 Experimental setup for the SCA-aided DFA procedure

The laser activation length was set to 150 ns and the laser power was set to 3%, resulting to ≈0.24 W.

As the device under test (DUT), we used ATmega328P microcontroller, decapsulated from the back-side and mounted on Arduino UNO development board. The area of the chip is 3×3 mm^2, while the sensitive area covers ≈1% of the whole chip size. The implementation of PRESENT we used computes the addRoundKey byte-wise and S-Box nibble-wise, therefore, we had to take this into account during the fault model estimation. The attack has to be likewise adjusted if the S-Box is computed byte-wise.

To control the impact location, we used XYZ positioning table with the spatial precision of 0.05 µm. Timing precision is achieved by inserting the trigger at the start of round 28.

For the side-channel leakage measurement, we used a digital oscilloscope, capturing the time frame of one round after the fault was injected. In order to distinguish the fault mask, we first profiled the standard power consumption by calculating an average of 100 encryptions. Then, another 100 experiments were done while injecting the fault at the same position and same timing and again, averaging these traces. Afterwards, we calculated a difference of these two traces and it gave us the knowledge of the injected fault mask in the previous round.

We would like to point out that both the triggering for fault injection and the averaging of the power traces are optional. By observing the side-channel trace, it was straightforward to pin-point beginning and ending of each round and each operation (S-BoxLayer, pLayer, addRoundKey) had a unique side-channel signature (this can be seen in the upper part of Fig. 4.9). Similarly, the fault mask was recoverable from single trace.

4.4.2 Determination of Fault Mask

In the following, we will detail the process of estimating the fault mask, based on laser fault injection parameters, and side-channel leakage.

There are two steps of determining the faulty nibble and the mask in round 28:

1. In order to get the information on which nibble has been faulted, we have to check the timing from the trigger. The S-Box computation on the DUT takes $\approx 11\ \mu s$. During the profiling phase, the timing for processing each nibble can be estimated with 100% success rate. Nibbles are processed in a reversed order (15 to 0), therefore, for example, w.r.t defined trigger signal, nibble 14 starts with a timing offset of $2,331$ ns, while nibble 0 starts at $12,537$ ns.
2. For estimating the fault mask, we have to check the side-channel leakage in the round 29. This process is depicted in Fig. 4.6 and the values corresponding to each power trace are detailed in Table 4.3. As can be easily seen, the difference trace shows the nibble position in round 29 which has a different value from the trace corresponding to non-faulty execution. This position can be easily determined by correlating the result with the profiling phase and comparing different traces. In Fig. 4.6, the red guiding lines show the relative position of nibbles, while '1' indicates that there is a difference w.r.t. the original trace. Then, by applying a reverse bit-permutation, the output S-Box difference of round 28 can be determined.

4.4.3 Key Recovery: Performance and Efficiency

We now present results illustrating the number of necessary fault injections to recover the last round key of PRESENT using the faults. Our SCA experiments demonstrated that nearly all the output fault masks obtained upon fault injection had Hamming weights of 1, 2 and 3, which is precisely the useful class of faults for the attack. Figure 4.7 presents a comparison of the estimated number of fault injections required under each kind of fault mask for fault propagation to a given number of nibbles in the ciphertext. Recall that the DFA can recover a given key nibble only if the fault propagates to that nibble in the last round. Hence, the efficiency of the attack depends on the average number of fault injections required for the fault to propagate to each nibble. Quite intuitively, greater the Hamming weight of the output fault mask in the target round, the faster the fault propagates to a larger number of nibbles in the subsequent rounds. This is also reflected in our experimental results. Next, we present the average number of fault injections required to recover a given number of nibbles of the last round key for each category of fault mask in Fig. 4.8. Quite evidently, key recovery for fault masks with Hamming weights 1 and 2 requires a significantly lesser number of fault injections than with Hamming weight 3. This is in accordance with the theoretical estimate for the required number of fault injections in Sect. 4.3.5. On an average, a combination of fault masks with Hamming Weights 1 and 2 recovers

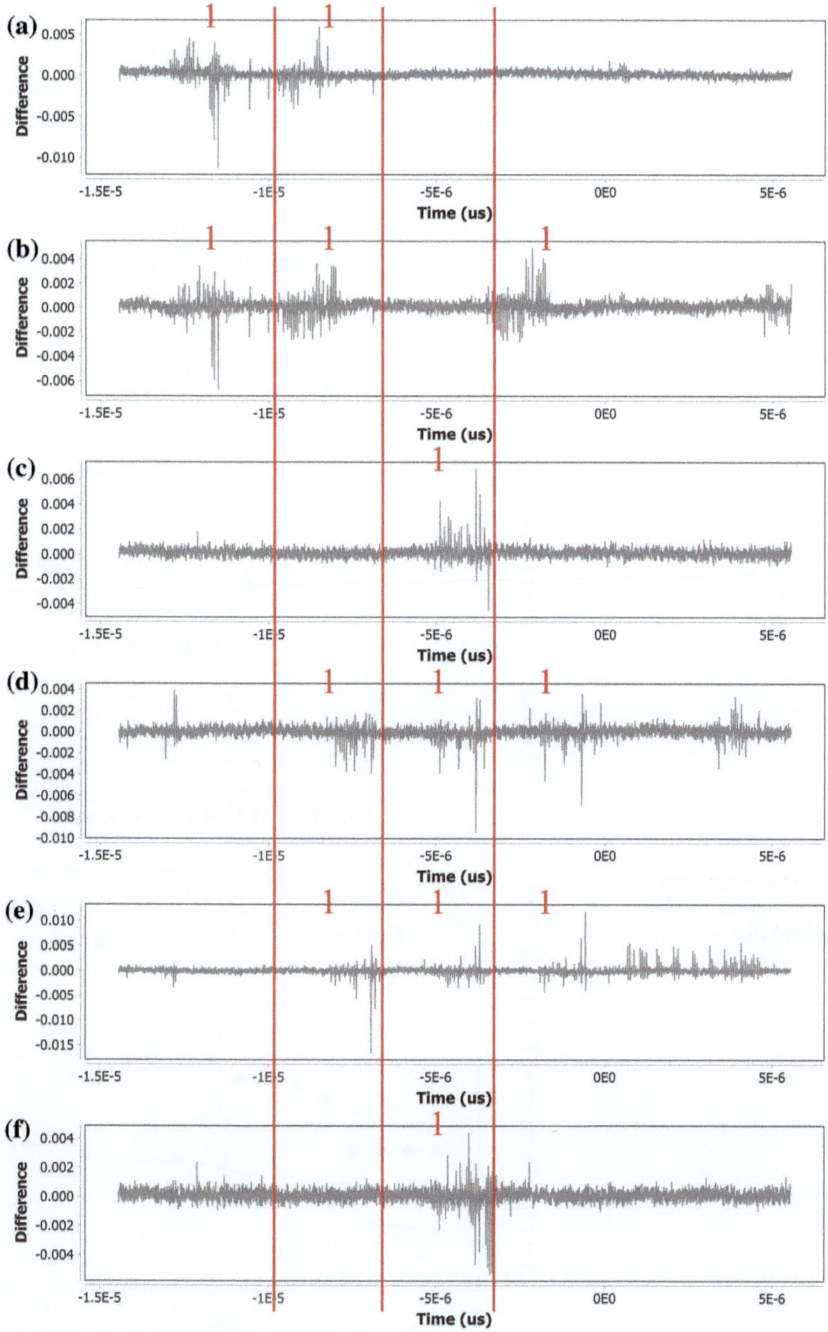

Fig. 4.6 Fault mask determination from SCA leakage

Table 4.3 Analysis of the leakage difference patterns from Fig. 4.6.

Trace	Offset (ns)	I/P Fault Mask:R29	O/P Fault Mask:R28
(a)	4032	0000000000800080	00000000000C0000
(b)	4914	0040000000400040	0000000000D00000
(c)	7686	0000080000000000	0000000200000000
(d)	9072	0200020002000000	0000070000000000
(e)	10143	0100010001000000	0000700000000000
(f)	10773	0000800000000000	0002000000000000

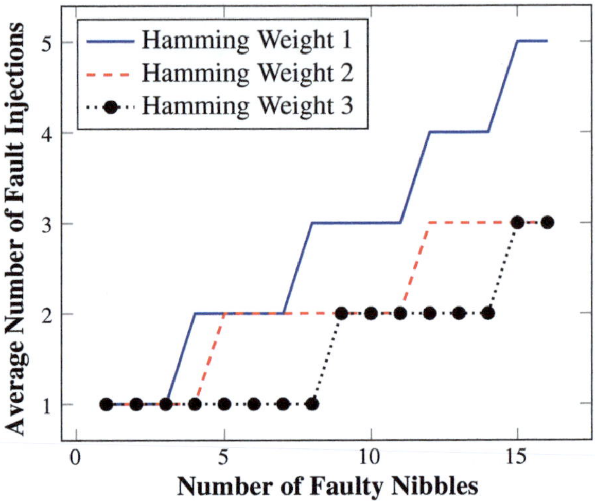

Fig. 4.7 Fault propagation: Average number of fault injections v/s Number of faulty nibbles

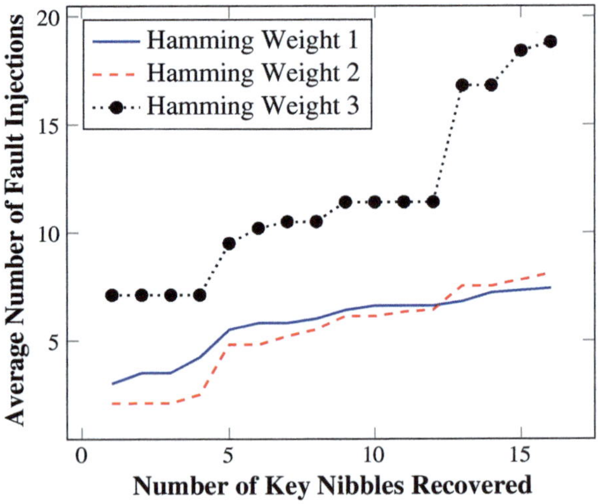

Fig. 4.8 Key recovery: Average number of fault injections v/s Number of key nibbles recovered

64 bits of the key in **7–8 fault injections**, which is at par with the best known prior fault attack on PRESENT, while the best case scenario allows recovering 64 bits of the key with 4 **fault injections**. The best case scenario is typically encountered when using fault masks of Hamming weight 1, and the fault diffuses to all nibbles of the final round in each injection. The worst case scenario, on the other hand, is encountered with fault masks of Hamming weight 3, when 19 fault injections are found to be necessary for key recovery.

4.5 Discussion

4.5.1 Extension to Other Rounds & Targets

While it is relatively easy to determine the faulty nibbles in round 29, this process becomes harder once the propagation of the fault produces collisions. For example, let us assume that after several rounds, the difference reaches nibbles 8 and 11, producing the S-Box output masks of value '8' for both nibbles. These two bits will propagate to nibble 15 in the next round, producing the S-Box input mask 'C'. Such a difference cannot be easily seen from the power trace by just observing the differential peaks, and therefore, one can only assume that some of the nibbles between 8 and 11 in the previous round were faulted. This behavior would require creation of SCA templates for each nibble and each fault mask, resulting in total of 256 different templates.

Similar scenario can be seen in Fig. 4.9, depicting the last three rounds of encryption—upper part is the power trace of the non-faulty encryption process, so that it makes it possible to estimate particular rounds. Lower part is the differential trace which shows that while it is trivial to determine the fault mask for round 28, following rounds are hard to analyze.

4.5.2 Extension to Other PRESENT-Like Block Ciphers

The combined SCA and FA based attack on PRESENT can also be extended to other PRESENT-like block ciphers that use bit permutations for diffusion. One such recently proposed block cipher is GIFT [16]. GIFT has two versions—GIFT-64 with a 64 bit plaintext block and GIFT-128 with a 128 bit plaintext block. The bit-permutation layer for GIFT-64 differs from PRESENT, which lends it some additional advantages against cryptanalytic attacks. However, similar to PRESENT, the bit-permutation layer of GIFT-64 also ensures that each of the four sets of nibbles, namely [0–4],[5–8], [6–11] and [12–15] affect exactly four non-overlapping sets of nibbles $\{0, 4, 8, 12\}$, $\{1, 5, 9, 13\}$, $\{2, 6, 10, 14\}$ and $\{3, 7, 11, 15\}$, respectively. Since it is this property of the bit-permutation that our combined attack exploits, the

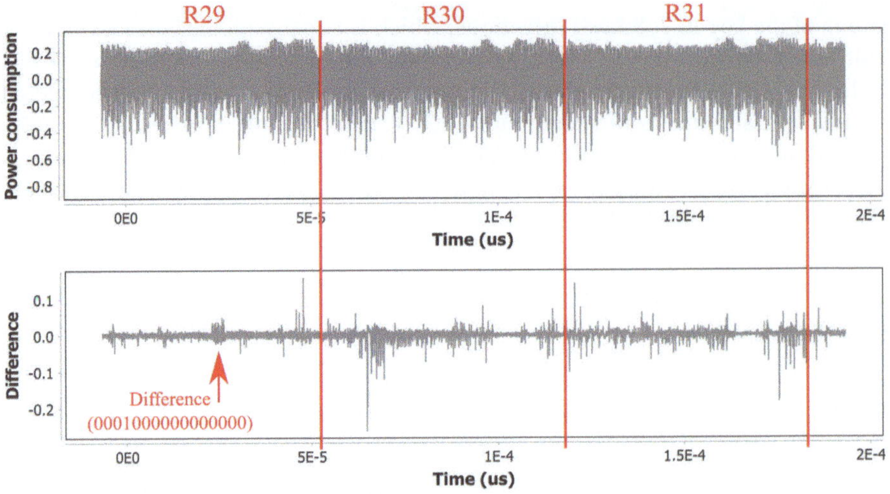

Fig. 4.9 Power trace for the last three rounds with a corresponding differential trace

attack is also applicable to GLITCH-64 with the same efficiency as PRESENT. An extension of the attack is also applicable in case of GLITCH-128, although a detailed description of the same is beyond the scope of the current chapter.

It is, however, interesting to see that the combined attack methodology would fail against ciphers such as AES or LED that use MDS layers. The diffusion characteristics of an MDS layer would break any correlation between the output fault mask of a prior round and the eventual input fault mask at a later round, by causing the fault to always diffuse to the maximum possible number of nibbles in each round. This leads to the following interesting observation: *While bit-permutations are a popular choice for lightweight block ciphers owing to their greater efficiency in hardware as compared to MDS layers, they are prone to a class of combined SCA+FA attacks that MDS layers resist*. To the best of our knowledge, no prior fault attack has shed light on this apparently inherent vulnerability of bit-permutation based diffusion layers towards combined attacks. A possible future work is to compare and contrast the vulnerability of MDS layers and bit-permutations against a wider class of fault attacks in general.

4.5.3 Possible Countermeasures

We briefly discuss potential countermeasures and their effectiveness in resisting our combined attack methodology on bit-permutation based SPN block ciphers. Standard fault detection mechanisms such as spatial and temporal redundancy could potentially increase the number of fault injections required; but they can be bypassed using precise and localized fault injection techniques such as laser beams that allow injecting the same fault in both the original and redundant computations with high probability. Standard side-channel countermeasures such as masking could be incorporated to make the attack more difficult in the sense that one might require higher order analysis over a larger number of power traces to retrieve the desired fault mask value in such a scenario. Other less costly alternatives to masking such as shuffling and hiding might also increase the attack complexity when retrieving the fault mask. It would be an interesting future work to quantify the exact security afforded by the presence of SCA countermeasures against the attack.

A final point to note is that although we experimentally demonstrate the attack on a software implementation of PRESENT, the attack is equally applicable on a hardware implementation as well. A hardware implementation processing the nibbles in a serial fashion, are equally vulnerable to such attacks. For parallel implementations, the adversary would need side-channel equipment with localized measurement capability to recover the fault mask. Once the fault mask is recovered, the same attack follows.

4.6 Summary

In this chapter, we have practically demonstrated the strength of combining side channel analysis (SCA) and fault analysis (FA) in attacking PRESENT and PRESENT-like block ciphers that use bit-permutations instead of MDS layers for diffusion. We use a laser fault injection based setup to inject nibble faults in the 28th round of PRESENT, and perform a DPA of the correct and faulty executions to fully infer the resulting fault mask. We subsequently demonstrate a generalized DFA expanding across the last four rounds of PRESENT that allows reducing the entropy of the input fault masks for the last round, and recovering multiple key nibbles in parallel. We have practically corroborated our theoretical observations via actual fault injection and key recovery attacks on an ATmega328P microcontroller-based implementation of PRESENT-80. The attack exposes an interesting and hitherto unexplored vulnerability of PRESENT-like block ciphers using bit-permutations to achieve diffusion. While bit-permutations are highly desirable for efficient hardware implementations, they lead to fault diffusion trails that could be exploited via SCA-assisted fault attacks, as demonstrated in this chapter. An interesting future scope of research is to investigate the effect of side-channel countermeasures on the proposed combination of SCA and FA in this chapter.

Part III
Advanced Fault Analysis Techniques and Fault Analysis Automation

Chapter 5
Laser-Based Fault Injection on Microcontrollers

Jakub Breier, Dirmanto Jap and Chien-Ning Chen

Abstract Laser fault injection constitutes a powerful tool for a precise injection of faults into the device, allowing an adversary to carefully adjust timing and position on the chip. On the other hand, the cost of such equipment is high and the profiling time is non-negligible. In this chapter, we provide a theoretical background on laser fault injection, followed by practical evaluation of this technique on 8-bit microcontroller. We first profile the device to examine what fault models are possible and then we provide a case study on ChaCha family of stream ciphers.

5.1 Introduction

Fault injection techniques can differ greatly from many points of view – price, repeatability, precision, and user-friendliness. One of the most popular techniques for fault injection is using laser source as the main component for disturbing the integrated circuit (IC). Laser fault injection is considered a very powerful and precise technique, with a high degree of repeatability. Drawbacks of such a technique are its high price and the necessity of trained and experienced personnel to operate the injection device.

When performing laser fault injections into an integrated circuit, the target area has to be directly accessible. Therefore it is necessary to de-package the chip, usually enclosed in an epoxy package. This can be done either by using specific types of acids (necessary for the front-side de-packaging) or by mechanical grinding and milling the epoxy layers (possible only for the back-side of the chip). Both sides of the chip have different properties that require different laser wavelengths. Since it is not necessary to penetrate the silicon substrate from the front side, it is possible to

J. Breier (✉) · D. Jap
Physical Analysis and Cryptographic Engineering, Temasek Laboratories, Nanyang Technological University, Singapore, Singapore
e-mail: JBreier@ntu.edu.sg

C.-N. Chen
Independent Researcher (In Collaboration with NTU), Singapore, Singapore

S. Patranabis and D. Mukhopadhyay (eds.), *Fault Tolerant Architectures for Cryptography and Hardware Security*, Computer Architecture and Design Methodologies, https://doi.org/10.1007/978-981-10-1387-4_5

use the red (808 nm) or the green (532 nm) laser to make the fault injection. However, because of the absorption properties of the silicon [73], we need to use at least near-infrared (1064 nm) laser in order to make the attack possible from the back side of the chip.

5.2 Historical Development

The development of state-of-the-art laser fault injection techniques can be traced back to the seminal work by Skorobogatov and Anderson [167], where a camera flash and a laser pointer were used to set/reset bits in the SRAM cell. With inexpensive fault injection equipment and a decapsulation of the microcontroller from the front side, they exposed the vulnerability of CMOS attacks to optical attacks. This was followed by the demonstration of byte fault attacks on AES [64] via SRAM injections. The traditional bit-flip model was initially found to practically infeasible using laser fault injections [159] (only bit set-reset fault models could be realized); however subsequent injection results [158] that both bit-flips as well as bit set-reset fault models are feasible. The latter results, however, used a very large laser beam ($125 \times 125\ \mu m^2$).

The first prominent work on backside laser fault injection on a 90 nm microcontroller was performed by Courbon et al. [49], that achieved a byte set-reset fault model. This was followed by an attack on an AES implementation on a 130 nm microcontroller [50], where a Scanning Electron Microscope was used to identify the target flip-flops after opening the IC from the front side. Agoyan et al. [1] presented a DFA attack on AES implemented on a 350 nm microcontroller by performing multiple byte faults. They attacked the surface of the chip from the front side, aiming at the SRAM. They used a green laser beam ($\sim$532 nm) with 5.5 μm diameter, 20× magnifying objective lens and a positioning table with 0.1 μm precision. Finally, laser fault injections on ASIC-based AES implementations were investigated by Roscian et al. [158]. The authors used a relatively large laser beam spot (square spot $125 \times 125\ \mu m^2$) to successfully perform bit flips and bit sets/resets, including single bit faults. This enabled them to launch a DFA of AES-128.

5.3 The Physics of Laser Fault Injection

Ionization effect on transistors is a well-known phenomenon. The first observation of such effect was described in 1975, when flip-flop circuits of communication satellites were triggered by cosmic rays [31]. More specifically, the mechanism for the cosmic ray interaction was the charging of the base-emitter capacitance of a transistor to the turn-on voltage. Since then, a single-event upset (SEU) simulation by lasers has become a standard methodology for testing integrated circuits, providing an inexpensive and fast alternative to the accelerated ion beam technique [43]. Because

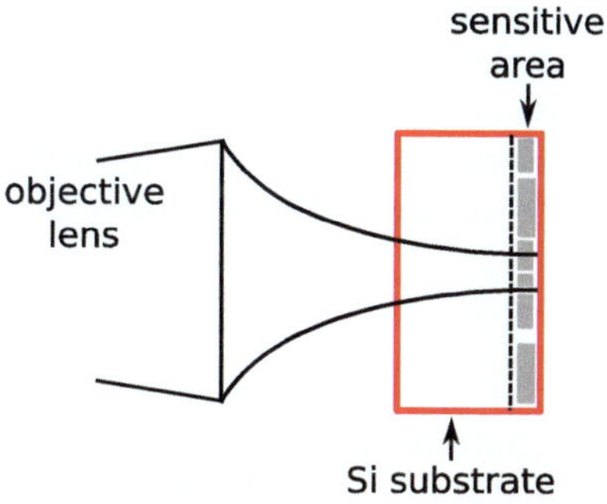

Fig. 5.1 Backside laser targeting [76]

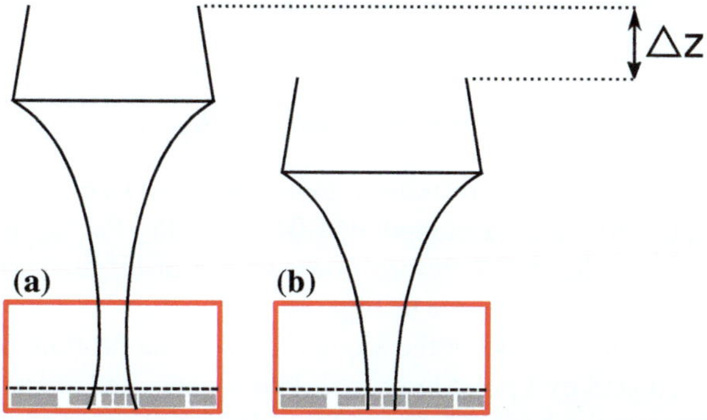

Fig. 5.2 Focusing on the sensitive spot – in case of (**a**), laser is focused on the backside surface of the chip, in case of (**b**), it is focused on sensitive components. In both cases, the effective laser spot will be different [76]

of the increasing number of the frontside metal layers, it was shown that the most effective way to disturb the circuits by the laser is by focusing the beam through the backside substrate [106]. When using this method, it is still necessary to focus the beam so that it affects the frontside of the chip where the components are positioned, as seen in Fig. 5.1. This figure also shows that it is important to have a constant thickness of the substrate, because any variation in wafer thickness results in different Z-position of the focal point with respect to sensitive components one wants to target. Also, the backside should be polished so that the beam quality would not be affected by silicon absorption and refraction.

As the beam passes through the surface, it converges. The angle of incidence varies across the beam from 0 degree to its maximum ($sin^{-1}(r_1/l)$), where r_1 is the lens radius and l is the distance of the lens from the surface [44]. This is also illustrated in Fig. 5.2. By changing the distance from the surface, the resulting spot size of the laser varies. First, the laser is focused on the surface (a), then it moves by distance Δz so that it is focused on sensitive components (b). This value can be calculated by the following equation:

$$\Delta z = \frac{t_{sub}}{\eta_{Si}} \tag{5.1}$$

where t_{sub} is the substrate thickness, and η_{Si} is the refractive index for silicon. For example, for near-infrared laser used for backside laser irradiation, the value is η_{Si} (1064 nm) = 3.5.

In order to determine the most effective wavelength for triggering SEU, the following formula gives the first-order approximation of the electron-hole pairs generation:

$$\frac{dN}{dt}(d) = \frac{\alpha\lambda}{hc} I(d) \tag{5.2}$$

where d is the substrate thickness, α is the absorption coefficient, λ is the wavelength, h is the Planc constant, and I is the laser intensity. Experimental measurements show that the following expression approximates the absorption in silicon [160]:

$$\alpha = \left(\frac{84.732}{\lambda} - 76417\right)^2 \tag{5.3}$$

It means, for a substrate thickness $d = 400\,\mu\text{m}$, the maximum generation rate is obtained for a wavelength of 1.04 μm [106]. For thinner substrates, optimum wavelength is lower. Therefore, lasers with near-infrared wavelengths are suitable candidates for the backside testing.

Figure 5.3 shows the situation when the NMOS drain of the OFF transistor is irradiated by a pulse laser [65]. Photo-generated holes (grey dots) flow to the ground line via the NMOS body and the photo-generated electrons (black dots) flow to the power line, via the PMOS transistor. However, if the laser is focused on the drain of the ON transistor, there will be no photocurrent generated and the transient voltage will not propagate to the gates on its output.

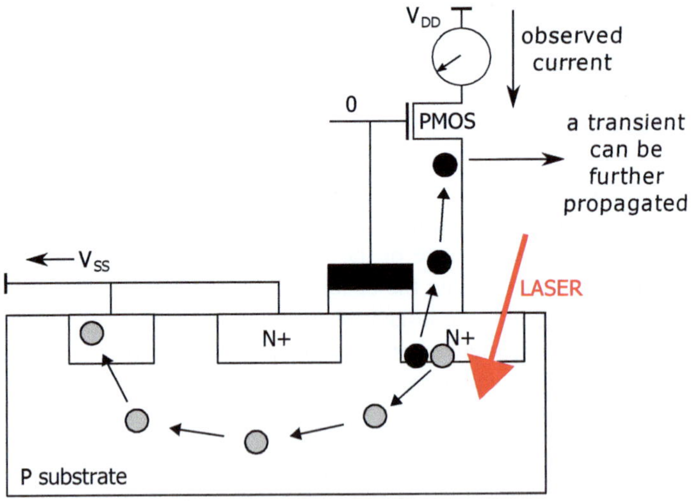

Fig. 5.3 Effect of laser when focused on the OFF transistor [76]

The effective spot size is another important parameter when targeting sub-micron components. It is mostly determined by the optics and the wavelength. If the laser spot is smaller than the component size, it is possible to determine the amount of charge deposited in the component by converting the laser energy to a linear energy transfer (LET) equivalent by the following formula [44]:

$$LET = (1 - R(\lambda)) \cdot \frac{E_0 \cdot E_p}{E_\gamma} \cdot \frac{1 - e^{\alpha(\lambda)d}}{\rho d} \tag{5.4}$$

where $R(\lambda)$ is the reflection coefficient, E_p is the electron-hole creation energy (3.6 eV/pair in Si), E_γ and E_0 are the photon and incident energies, respectively. ρ is the material density, $\alpha(\lambda)$ is the absorption coefficient, and d is the charge collection depth.

5.4 Device Decapsulation

Laser fault injection on a target device often requires depackaging the chip, so that the laser beam may approach the device surface. The process of depackaging must take into account two main factors - the chip package and also which side (front or back) of the silicon die surface the laser beam will approach. Based on these considerations, there are two main depackaging methods – *chemical decapsulation*, using acids to dissolve the epoxy layer covering the silicon die and *mechanical decapsulation*, using mechanical milling devices to reach the surface. We describe these techniques in the following discussion.

5.4.1 Chemical Techniques

Chemical decapsulation typically uses either a fuming nitric acid (HNO_3 with concentration $\geq$90%), a concentrated sulphuric acid (H_2SO_4 with concentration $\geq$96%), or a combination of these acids. It can be used for both frontside and backside decapsulation of a given chip. For example, the authors of [133] proposed that a mixture of 80% HNO_3 and 20% H_2SO_4 is a good option for decapsulating ICs with copper wire bonds, without harming the wiring. Since the chemical ingredients are mostly corrosive in nature, this technique requires a chemical laboratory with specialized equipment, including fuming hoods to avoid acid vapor inhalation. See, for example, Figs. 5.4 and 5.5 for a typical laboratory setting, and the corresponding chemicals used for decapsulation.

The process consists of following steps:

1. Heating the acid to 30–60 °C (optional step, makes the process faster).
2. Applying small portion of the acid on the epoxy surface.

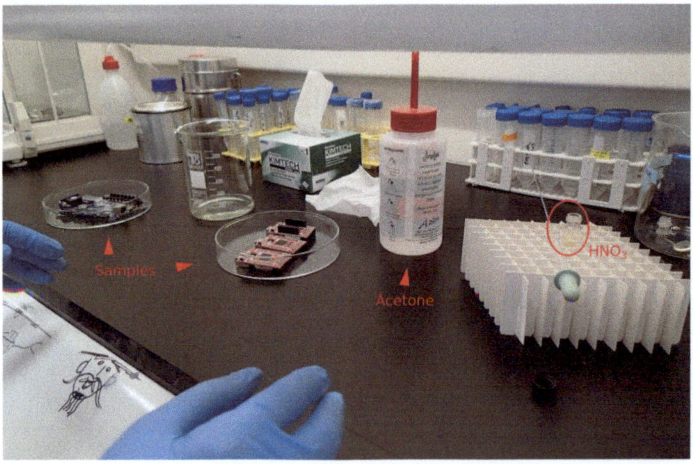

Fig. 5.4 Typical laboratory settings for chemical decapsulation [39]

Fig. 5.5 Detail of the samples [39]

3. Washing the acid with the acetone.
4. Repeating steps 2–3 until desired portion of the chip is visible (this step depends on the target area of the chip and the nature of the bonding wires in the chip).

As a rule of thumb, ICs with golden wires are the easiest to decapsulating since they do avoid the risk of damage, unless some wire contacts are exposed to mechanical cutting. Hence, the whole IC surface can be exposed to the acid while decapsulating. Figure 5.6 shows such a depackaged chip. ICs with copper wires, on the other hand, need to be carefully decapsulated by exposing only the central region of the die to the acid; otherwise, the acid could potentially dissolve all the copper wirings once it reaches the edge of the chip (see Fig. 5.6).

The advantages and disadvantages for chemical decapsulation are summarized in Table 5.1.

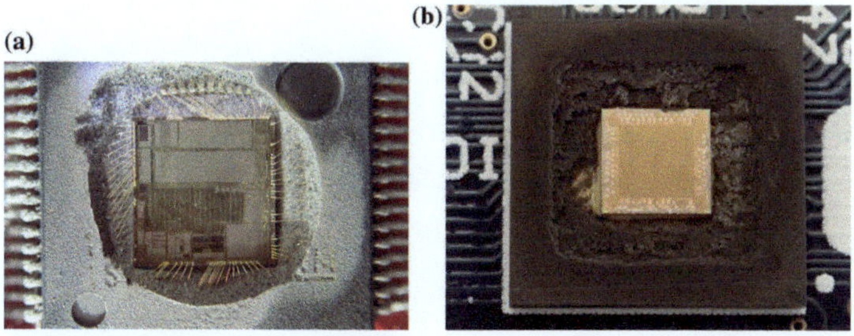

Fig. 5.6 Chemical decapsulation – successful (**a**), and unsuccessful (**b**) [39]

Table 5.1 Advantages and disadvantages of chemical decapsulation [39]

Advantages	Disadvantages
• Only technique for the frontside decapsulation	• Need to perform in a laboratory
• Does not require expensive equipment	• Fuming acid is a health hazard
• Does not leave mechanical damage on the surface	• Acid can dissolve bonding wires on the edge of the chip, making it unusable
• Relatively fast	• Low precision
	• Uneven delayering

5.4.2 *Mechanical Techniques*

Unlike chemical decapsulation, mechanical decapsulation cannot be applied to the frontside of an IC, since the milling head would almost certainly destroy the IC components on reaching them. It could in fact destroy the entire chip from the excessive heat generated before the process. But a precise milling equipment can be used not only to remove the epoxy package of the device, but also to delayer the silicon substrate. This is especially beneficial since the effectiveness of the laser fault injection depends to a large extent on the thickness of the silicon substrate. As mentioned in [72], the effective penetration of a laser beam varies for various wavelengths - approximately 1.58 μm for 532 nm wavelength, approximately 12.79 μm for 808 nm wavelength, and approximately 1100 μm for 1064 nm wavelength.

Figure 5.7 demonstrates a basic decapsulation of a standard low-cost AVR microcontroller decapsulated using a Proxxon precision drill/grinder. The milling is performed in several passes to help the chip cool down in the interim. Removal of the epoxy layer is followed by thinning the copper substrate until it can be peeled off easily. The final layer of glue holding the silicon die is removed by scratching off using hard plastic tools. Alternatively, high-end chips could be ball grid array (BGA) flip chip package, which has a metal covering instead of an epoxy one. The metal covering is typically held on to the chip surface using glue, which can be heated to easily remove the covering. However, the silicon substrate is typically much thicker

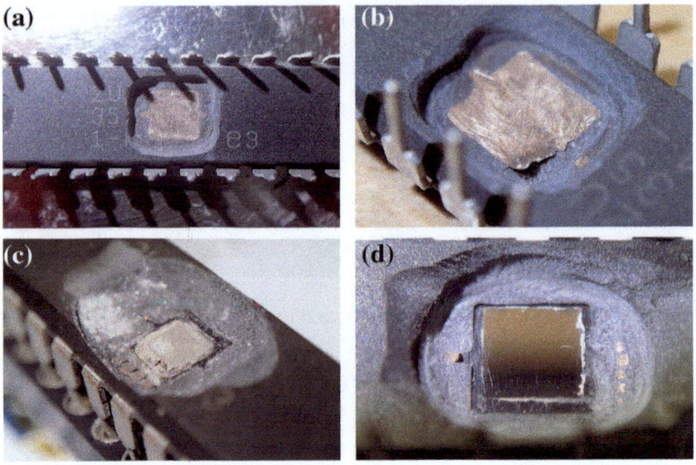

Fig. 5.7 Mechanical decapsulation steps: **a** grinding, **b** thinning the copper substrate, **c** removing the substrate, **d** removing the glue and polishing [39]

Table 5.2 Advantages and disadvantages of mechanical decapsulation [39]

Advantages	Disadvantages
• Provides high precision	• Cannot be used for the frontside
• Can be used for delayering of the silicon substrate	• Excessive heat can damage the chip
• Can be automated	• Precise equipment is expensive
• Does not have to be performed in a laboratory	

in such devices, which makes optical fault injection difficult. High precision equipment such as Ultra Tec ASAP-1 and prior sample preparation are required in such cases to thin the substrate layer carefully.

The advantages and disadvantages for mechanical decapsulation are summarized in Table 5.2.

5.5 Targeting a Microcontroller

When actually targeting a microcontroller, it is vital to choose a laser source of appropriate wavelength in order to achieve effective penetration. For instance, when targeting the front side, the IC components are directly visible after decapsulation, and one need not penetrate the silicon substrate in such cases. Consequently, a 532 nm (green) or an 808 nm (red) pulse can be used. For backside attacks, on the other hand, high penetrability is desired; hence, 1064 nm (near infrared) pulses are preferable. Such longer wavelengths typically have penetration capabilities more than 1100 μm. However, short wavelength pulses possess more energy, and produce faults with greater accuracy.

The next parameter to fix is the laser spot size. State-of-the-art manufacturing technologies allow for very small transistor sizes, which necessitates small spot sizes for precise injections. It has been observed that one can attack several technologies with laser injections, ranging from ranging from low-end microcontrollers manufactured by 250 nm process, to high-end FPGAs with transistor size 65 nm. These devices can be subject to bit flips with high success rate($\geq$90%) using laser spots varying between 10–800 μm^2. Authors in [158] claim they were able to disturb particular bits even with a very large laser beam spot ($125 \times 125\ \mu m^2$), although the success rate was lower.

The final parameter to take care of is the laser power. The preferable strategy is to gradually increase the power level while keeping the laser activation period short. This ensures that the beam is not activated for a very long period of time, since this might lead to loss of precision. The near-infrared laser beam documented in this chapter has a maximum output power of 20 W, which in turn reduced to 7–8 W via objective lenses. It is observed that perturbing smaller size technology requires higher energy pulses. For example, while 65 nm FPGAs require the laser power had to be at least 80% in order to produce fault injections, laser powers between 10–20% suffice for older microcontrollers.

5.6 A Case Study of Laser-Based Fault Attacks on the ChaCha Family of Stream Ciphers

One of the most popular form of fault attacks using laser pulse injections onto microcontrollers is the instruction skip attack. In this attack, the adversary alters the intended flow/operation sequence of a given algorithm by injecting a fault just when a given instruction is about to be executed, thereby corrupting the instruction register. There are two possible outcomes of such an attack - either the instruction gets entirely skipped, or gets replaced by an alternative instruction. Such fault attacks are often devastating to the security of software-based cryptographic implementations. For example, think of a scenario where one could simply skip the last key XOR operation of AES-128. Such an attack would trivially reveal the last-round key to the adversary.

In this section, we present to the readers a case study of instruction skip/instruction replacement attacks using the ChaCha [24] family of addition-rotation-XOR (ARX) based stream ciphers as a target. The introduction of the Salsa family of stream ciphers by Bernstein in [25] has caused the ARX family of crypto-primitives to gain popularity in cryptographic literature. Besides Salsa, there exists today a number of block ciphers (e.g. SPECK [23]) and hash functions (e.g. BLAKE [11]) that use the ARX design paradigm. As the name suggests, ARX uses a combination of modular addition (for non-linearity), rotation and XOR operations, as opposed to the standard use of substitution-boxes (S-Boxes) along with linear operations. An ARX operation is typically of the form $d = ((a + c) \lll k) \oplus b)$ where k is usually a constant, and a, b, c and d are registers of appropriate width. ARX operations usually afford lower diffusion per round; consequently ARX-based crypto-primitives usually require a larger number of rounds. However, ARX-based designs avoid the need for S-Box lookups, which makes them efficient in software and resistant to timing-based side-channel attacks.

5.6.1 *The ChaCha Family of Stream Ciphers*

The ChaCha family of stream ciphers was proposed by Bernstein in [24]. It is an improvement over the original Salsa family of stream ciphers, with increased levels of diffusion per round. The basic operational unit for ChaCha is 32-bit words. The *ChaCha Function* maps a 256-bit key $k = \{k_0, k_1, ..., k_7\}$, a 64-bit nonce $v = (v_0, v_1)$, and a 64-bit counter $t = (t_0, t_1)$ to a 512-bit keystream block. In particular, this function takes as input the 4×4 matrix of 32-bit words written as:

$$X = \begin{pmatrix} x_0 & x_1 & x_2 & x_3 \\ x_4 & x_5 & x_6 & x_7 \\ x_8 & x_9 & x_{10} & x_{11} \\ x_{12} & x_{13} & x_{14} & x_{15} \end{pmatrix} = \begin{pmatrix} c_0 & c_1 & c_2 & c_3 \\ k_0 & k_1 & k_2 & k_3 \\ k_4 & k_5 & k_6 & k_7 \\ t_0 & t_1 & v_0 & v_1 \end{pmatrix} \tag{5.5}$$

where c_0, c_1, c_2 and c_3 are the predefined constants $\{0 \times 61707865, 0 \times 3320646E, 0 \times 79622D32, 0 \times 6B206574\}$ respectively, and outputs a 512-bit keystream Z. The overall ChaCha function comprises of 20 rounds, where each round function is based on the following nonlinear operation (also called the *quarterround* function), which transforms a vector (x_0, x_1, x_2, x_3) to (y_0, y_1, y_2, y_3) by sequentially computing (Table 5.3):

$$\begin{aligned} b_0 &= x_0 + x_1, b_3 = (x_3 \oplus b_0) \lll 16 \\ b_2 &= x_2 + b_3, b_1 = (x_1 \oplus b_2) \lll 12 \\ y_0 &= b_0 + b_1, y_3 = (b_3 \oplus y_0) \lll 8 \\ y_2 &= b_2 + y_3, y_1 = (b_1 \oplus y_2) \lll 7 \end{aligned} \tag{5.6}$$

Table 5.3 The ChaCha quarter round function

Round	ARX input vectors to the round function
Odd numbered	$(x_0, x_5, x_{10}, x_{15}), (x_1, x_6, x_{11}, x_{12}), (x_2, x_7, x_8, x_{13}), (x_3, x_4, x_9, x_{14})$
Even numbered	$(x_0, x_1, x_2, x_3), (x_4, x_5, x_6, x_7), (x_8, x_9, x_{10}, x_{11}), (x_{12}, x_{13}, x_{14}, x_{15})$

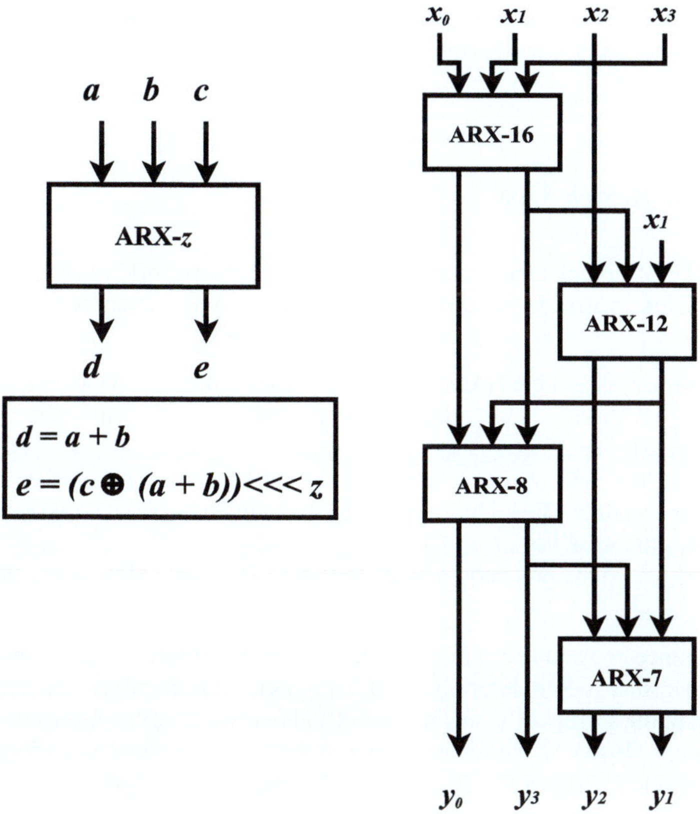

Fig. 5.8 The quarterround function of ChaCha

The quarterround function is applied to the state matrix in row-major and diagonal-major fashion in the even and odd numbered rows, respectively. Let X^r be the state matrix after round r. The final output keystream block Z is computed as $Z = X \boxplus X^{20}$, where $\boxplus$ denotes word-wise integer addition.

ChaCha is efficient in software and resistant to timing side-channels. ChaCha, along with its authentication counterpart Poly1305, currently features in a new TLS cipher suite in Google Chrome that operates three times faster than AES-GCM on devices that don't have AES hardware acceleration, including most Android phones, wearable devices such as Google Glass and older computers (Fig. 5.8).

In this section, we present four differential fault analysis (DFA) attacks on an microcontroller-based implementation of ChaCha using the instruction skip and instruction replacement fault models and attack methodologies introduced to the readers. The attacks showcased in this section target the keystream generation module of ChaCha at the decryption site. The attacks are practically demonstrated on an Atmel AVR 8-bit microcontroller-based implementation of ChaCha using a laser fault injection setup. Each of these attacks using instruction skips requires around

5–8 fault injections on an average to recover the entire 256 bit secret key, while the attack using instruction replacements requires **32 fault injections** on an average for full key recovery.

5.6.2 The Attack Model

For the sake of completeness and ease of the readers, we briefly recall the attack model for DFA on stream ciphers in Fig. 5.9. The attack steps may be enumerated as follows:

1. The adversary is assumed to *know* both the plaintext message M and the ciphertext C during encryption. This obviously implies that the adversary also knows the correct keystream K.
2. The target for fault injection is the equivalent keystream generation algorithm during decryption. The adversary obtains the faulty message M' and recovers the faulty keystream $K' = M' \oplus C$.
3. The adversary now proceeds with the standard DFA procedure using the knowledge of K and K'.

Since the nonce and counter values are identical during both encryption and decryption, nonce misuse is completely avoided. Thus, as long as the adversary can obtain at least a part of the secret key from the correct and faulty pair of output keystreams, the whole key can always be recovered from multiple fault injections targeting different parts of the key.

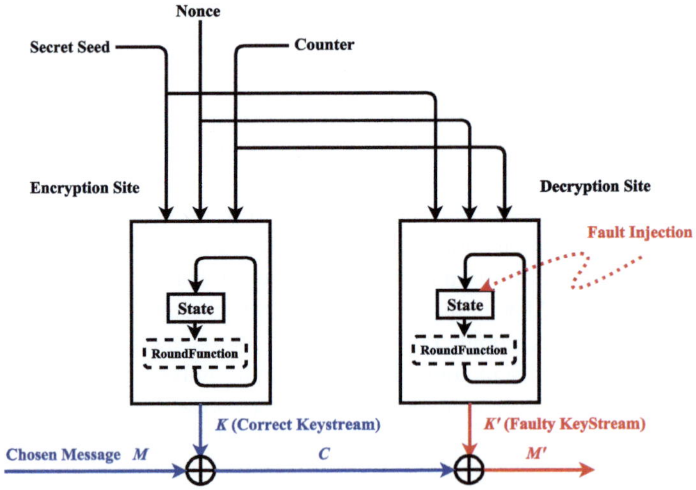

Fig. 5.9 DFA on stream ciphers: attack model without nonce misuse

5.6.3 The Target Implementation of ChaCha

We target an Atmel AVR microcontroller-based implementation of ChaCha. The ISA for this microcontroller comprises of standard instructions such as `add`, `sub`, `mov` and `cmp`. The microcontroller uses a 16-bit address-space, while being an 8-bit processor. Consequently, most of the instructions use 8-bit operands. Our implementation uses the general-purpose register set ($r16, r17, \ldots, r25$), and the additional register set ($r26, r27, \ldots, r31$) for indirect addressing.

5.6.3.1 Implementation of the Quarterround Function

We present snippets of the AVR code for implementing the quarterround function in Listing 5.1. The register $r19$ stores the input vector for the quarterround operation. An example comparison step with $r19$ is illustrated in Line 2. If $r19$ stores 0x00, then the first column comprising of x_0, x_4, x_8, x_{12} is treated as the input, and the instructions between Line 4 and Label 80 are executed. On the other hand, if $r19$ does not hold 0×00, the control branches to Label 40, followed by another immediate branch to Label 80. The process repeats iteratively until a matching condition for $r19$ is encountered. The use of two conditional branch instructions is motivated by the limitation on the range of addresses that a single jump can perform in the Atmel ISA. The input vector indices to the quarterround function for different values of the register $r19$ are depicted in Table 5.4. The fault attacks presented in this section target one or more instructions using a laser fault injection setup, as described next. The aim of injection is to skip/alter the target instructions, resulting in alterations to the control flow of the original program, which are subsequently exploited to efficiently retrieve the key.

5.6.3.2 Laser-Based Experimental Setup and Device Profiling

In this section we first explain the experimental setup which was used for the fault injection. Later, we will detail the attack parameters that were determined during the profiling phase.

Table 5.4 Inputs to the quarterround function

Content of $r19$	0×00	0×01	0×02	0×03	0×04	0×05	0×06	0×07
Input vector indices to quarter-round	(0, 4, 8, 12)	(1, 5, 9, 13)	(2, 6, 10, 14)	(3, 7, 11, 15)	(0, 5, 10, 15)	(1, 6, 11, 12)	(2, 7, 8, 13)	(3, 4, 9, 14)

Listing 5.1 Subroutine *ARX* : Implementation of quarterround function

```
ARX%=:
cpi r19, 0x00
brne 40f
/∗ Column 1 : (0,4,8,12)∗/
ldi r27,hi8(Array)
ldi r26,lo8(Array)
ldi r29,hi8(State + 0x00)
ldi r28,lo8(State + 0x00)
ldi r31,hi8(State + 0x10)
ldi r30,lo8(State + 0x10)
/∗Relative call to Plus32∗/
rcall Plus32%=
ldi r29,hi8(State + 0x30)
ldi r28,lo8(State + 0x30)
ldi r31,hi8(State + 0x00)
ldi r30,lo8(State + 0x00)
/∗Relative call to Xor32∗/
rcall Xor32%=
ldi r29,hi8(State + 0x30)
ldi r28,lo8(State + 0x30)
ldi r31,hi8(State + 0x30)
ldi r30,lo8(State + 0x30)
/∗Relative call to Rotate32∗/
ldi r22, 0x10
rcall Rotate32%=
...
...
40:
cpi r19, 0x00
brne 80f
...
...
80:
cpi r19, 0x01
brne 41f
...
...
/∗Rest of the code∗/
```

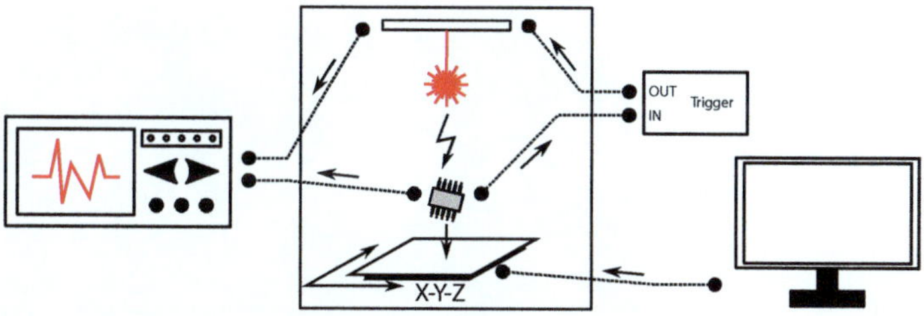

Fig. 5.10 Experimental setup for laser fault injection

5.6.3.3 Setup Components

The components of the setup and connections between them are depicted in Fig. 5.10. We describe the various components of the fault injection setup below:

- **Device under test**. As the device under test (DUT), we have selected a general-purpose 8-bit microcontroller, Atmel ATmega328P. It operates at 16 MHz, therefore one clock cycle takes 62.5 ns. The area of the chip is 3×3 mm^2 large. Before the experiment, the DUT was decapsulated by mechanical milling tools and polished by using Ultra-TEC ASAP-1 sample preparation equipment, in order to provide enough precision for the laser beam. This chip was mounted on the Arduino UNO board, adjusted for the laser testing purposes. The board communicates with the PC using the USBCDC interface. Figure 5.11 shows picture of the DUT during the experiment.
  Code for experiments was written in assembly language, by using Arduino programming framework. We set a trigger signal on the board to HIGH (5 V) before performing the operations in order to correctly identify the desired time. The board was mounted on an X-Y-Z positioning table with the step precision of 0.05 μm.
- **Optical source**. We utilized a near-infrared diode pulse laser (1064 nm) with maximal pulse power 20 W. The power was further reduced to 8 W by using a 20× magnifying objective lens. Laser spot size with this lens is 15×3.5 μm^2 and response to trigger pulse is approximately 100 ns.

A trigger device was used for capturing the trigger signal from the device and adjusting the laser activation timing. A digital sampling oscilloscope was used for verifying the laser pulse activation time and for estimating the timing of sub-operations in the cipher implementation. Finally, a PC workstation was used for communication with the DUT and analysis of results.

5.6.3.4 Laser Fault Injection Profiling

To profile the device, we have to vary several parameters in order to determine the optimal setup settings:

Fig. 5.11 Device under test during the experiment

- **Location of the laser spot**. Since we did not have a precise information about sensitive region of the chip prior to the experiment, we had to determine it experimentally. We scanned the entire region of the chip and estimated the area of interest, which was $50 \times 80\ \mu m^2$, corresponding to approximately 0.044% of the whole chip area.
- **Laser source power**. It is necessary to set the optimal laser power so that it is below destructive threshold, but it is high enough to disturb the execution. We were able to get the optimal results with the power of 4% of the total output power.
- **Timing offset from the trigger**. Distinguishing the right timing for the laser activation can be done empirically as well. In case the trigger position is known, one can just calculate the timing of target instruction by counting the clock cycles and adding the propagation delay of 100 ns. However, it is possible to determine the timing by simply observing the faulty ciphertexts and checking which part was changed. This will provide the information as to which part of the algorithm was affected.
- **Laser activation length**. This parameter has two distinct outcomes. It controls the overall energy that affects the device and therefore, if it is too short, faults will not appear. Also, it determines number of instructions that are disturbed in case ALU is targeted for the fault attack.

We now present four DFA attacks on ChaCha using the instruction replacement faults injected with the help of laser fault injection setup described above. The faults target the implementation of the ARX quarterround, described in Listing 5.1, as well as the final addition between the initial state X and the output, X^{20} of round 20 of the keystream generation algorithm.

5.6.4 An Instruction Skip Attack on the Final Addition Operation

5.6.4.1 Attack Description

The first DFA attack targets the final word-wise addition of the initial state containing the constants, key, counter and nonce to the final state generated after round 20. The attack skips one of the 16 word-additions, causing the faulty keystream output to trivially reveal the corresponding word of the secret key. The DFA procedure for recovering the key word k_1 is depicted in Fig. 5.12. Let Z and Z' be the correct and faulty keystreams respectively. Then, we have:

$$
\begin{aligned}
z_{1,1} &= k_1 + s_{1,1}^{20} \\
z_{1,1}' &= s_{1,1}^{20} \\
k_1 &= z_{1,1} - z_{1,1}'
\end{aligned} \tag{5.7}
$$

Solving the set of Equations in (5.7), we recover the key word k_1. Note that subtraction is performed *modulo* 2^{32}. The attack may now be repeated on each of the 8 words in the second and third row of the keystream, to retrieve each of the 8 words of the secret key.

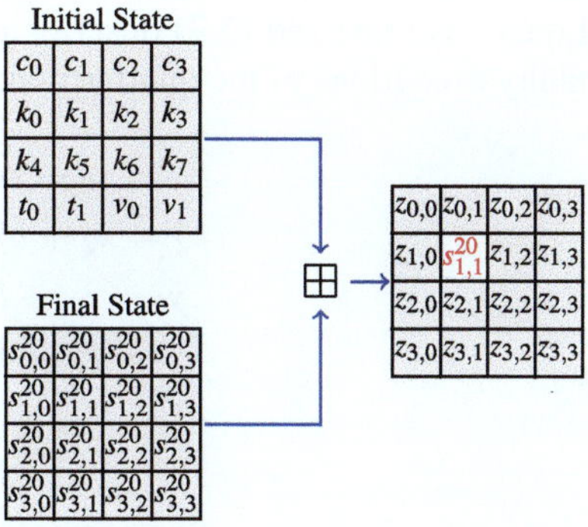

Fig. 5.12 First instruction skip attack: recovering k_1

5.6.4.2 Attack Realization

We now discuss a practical fault injection setting to mount the above attack on our AVR-based implementation of ChaCha. In the implementation, the 16-bit addresses of the words to be added are loaded in indirect addressing registers, and a 32-bit addition subroutine is invoked. In order to inject the desired fault in our laser-based setup, we set the trigger to just before the commencement of the final addition operation. The injection successfully skipped the function call to the 32-bit addition subroutine, and trivially revealed the corresponding word in the secret key. The fault injection was found to be 100% repeatable for each of the 8 potential target words, with varying injection timings.

5.6.5 An Instruction Skip Attack on a Rotation Operation

In any ARX-based cipher, the intermediate state is circular-rotated by a certain number of bits at some stage during quarterround execution. In ChaCha, for example, the various intermediate state registers are rotated by an offset of 16, 12, 8 and 7 bits in each quarterround. In our second attack, we target one such rotation in each quarterround of round 20 in the key generation algorithm. We inject faults to perturb the rotation offset, and exploit the resulting keystream via a DFA to recover two words of the secret key.

5.6.5.1 Attack Description

We describe a fault attack that targets the 12-bit circular rotation in each quarterround of round 20 in the key generation algorithm. The attack essentially skips a single-bit circular shift, resulting in a 11-bit overall rotation instead of the expected 12-bit rotation. Now, let Z and Z' be the correct and faulty keystreams, respectively. Equations (5.8) and (5.9) describe the quarterround operations for the correct and faulty executions of the quarterround, respectively.

$$\begin{aligned} b_0 &= x_0 + x_1, b_3 = (x_3 \oplus b_0) \lll 16 \\ b_2 &= x_2 + b_3, b_1 = (x_1 \oplus b_2) \lll 12 \\ y_0 &= b_0 + b_1, y_3 = (b_3 \oplus y_0) \lll 8 \\ y_2 &= b_2 + y_3, y_1 = (b_1 \oplus y_2) \lll 7 \end{aligned} \tag{5.8}$$

$$\begin{aligned} b_0 &= x_0 + x_1, b_3 = (x_3 \oplus b_0) \lll 16 \\ b_2 &= x_2 + b_3, b_1' = (x_1 \oplus b_2) \lll 11 \\ y_0' &= b_0 + b_1', y_3' = (b_3 \oplus y_0') \lll 8 \\ y_2' &= b_2 + y_3', y_1' = (b_1' \oplus y_2') \lll 7 \end{aligned} \tag{5.9}$$

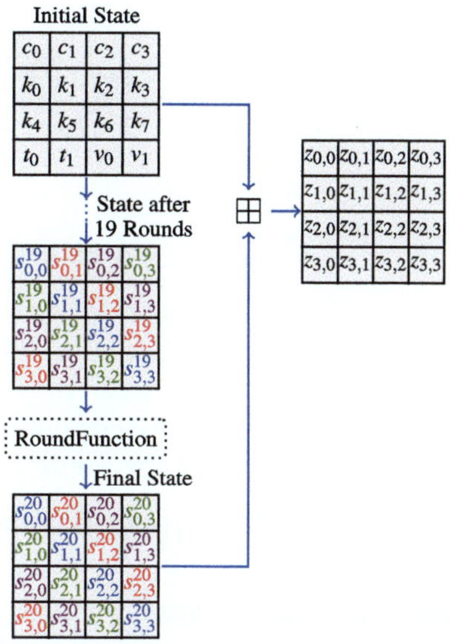

Fig. 5.13 Second instruction skip attack: perturbing rotation offset

The inputs x_0, x_1, x_2, x_3 in Eqs. (5.8) and (5.9) are the main diagonal words of the state after round 19, (see $\{s^{19}_{0,0}, s^{19}_{1,1}, s^{19}_{2,2}, s^{19}_{3,3}\}$ in Fig. 5.13). Similarly, the outputs $\{y_0, y_1, y_2, y_3\}$ and $\{y'_0, y'_1, y'_2, y'_3\}$ are the main diagonal words after *Round* 20, $\{s^{20}_{0,0}, s^{20}_{1,1}, s^{20}_{2,2}, s^{20}_{3,3}\}$. Our aim is to recover the final state words in the first diagonal, namely $\{y_0, y_1, y_2, y_3\}$, which then trivially reveal the corresponding words in the secret key.

Observe that after the completion of round 20, the first and fourth rows of final state are added to constants, counter values and nonce values, all of which are publicly available. This in turn reveals the following correct and faulty outputs of the quarterround: $\{y_0, y_3, y'_0, y'_3\}$ (see the system of equations below):

$$\begin{aligned} y_0 &= z_{0,0} - c_0 \\ y_3 &= z_{3,3} - v_1 \\ y'_0 &= z'_{0,0} - c_0 \\ y'_3 &= z'_{3,3} - v_1 \end{aligned} \tag{5.10}$$

Now, the difference between the correct and faulty keystreams reveal their difference in the respective final states after round 20, leading to the computation of both b_0 and b_1 as:

$$b_1 = b'_1 \lll 1$$

that is,

$$b_1' = \begin{cases} (b_1 + 2^{32} - 1)/2 & \text{if } \mathrm{LSB}(b_1) = 1 \\ b_1/2 & \text{otherwise} \end{cases}$$

$$\begin{aligned} b_1 - b_1' &= z_{0,0} - z_{0,0}' = y_0 - y_0' \\ y_2 - y_2' &= z_{2,2} - z_{2,2}' = y_3 - y_3' \\ y_1 - y_1' &= z_{1,1} - z_{1,1}' \end{aligned} \tag{5.11}$$

Substituting the values we computed from the set of Eq. (5.11), we just have y_2 left to solve for finding the key words, k_1 and k_6. Now, observe the following relation:

$$y_1 - y_1' = ((b_1 \oplus y_2) \lll 7) - ((b_1' \oplus (y_2 - z_{2,2} + z_{2,2}') \lll 7)$$

where y_2 is the only unknown. A simulation over all 2^{32} values of y_2 reveals that on an average, the above equation is satisfied by 2^{10} values of y_2, which in turn maps to 2^{10} unique pairs of values for (k_1, k_6) (see the following Eq. (5.12)).

$$\begin{aligned} y_1' &= (b_1' \oplus y_2') \lll 7 \\ y_1 &= (b_1 \oplus y_2) \lll 7 \\ k_1 &= z_{1,1} - y_1 = z_{1,1}' - y_1' \\ k_6 &= z_{2,2} - y_2 = z_{2,2}' - y_2' \end{aligned} \tag{5.12}$$

Thus the fault attack reduces the search space for the pair (k_1, k_6) from 2^{64} to 2^{10}. Repeating the same experiment for the three other diagonals finally reduces the search space of the entire 256-bit secret key from 2^{256} to around 2^{12}.

5.6.5.2 Attack Realization

In the AVR-based implementation of the quarterround, the attack on the rotation of the state registers is realized as follows:

- When the number of bits to be rotated is a multiple of 8, the circular rotation is a simple permutation of the 8-bit registers. In this case, there is no single-bit shift operation that could be skipped using the laser injection setup.
- However, when the number of bits to be rotated is *not* a multiple of 8, say 12, then the register permutation must be followed by four single-bit circular shifts, `LftShift` subroutine in Listing 5.2, one of which is targeted with 100% repeatability using the laser injection setup.

The AVR code for circular shift by 12 bits, and the corresponding instructions to be skipped are depicted in Listing 5.2. In particular, the laser trigger may be set just before round 20, with the $r19$ register holding the main diagonal entries to be passed to the quarterround as inputs.

Listing 5.2 Rotation by 12 bits

```
...
51:
ld r23, Y+
ld r24, Y
st Y+, r23
ld r25, Y
st Y+, r24
ld r23, Y
st Y, r25
st Z, r23
/*Single Bit Rotate Subroutine*/
rcall LftShift%=
rcall LftShift%=
rcall LftShift%=
rcall LftShift%=
rjmp RotateExit%=
...
```

5.6.6 A Diagonal Fault Attack via Alteration of Control Flow

Our third attack is a demonstration of a diagonal fault attack via a change in the control flow of the target program in round 20 of the keystream generation algorithm. Recall that Listing 5.1 depicted the use of branching instructions in the AVR code for the quarterround operation of ChaCha. In particular, observe in Fig. 5.14 that the operation on any diagonal involves a series of `cpi-brne` instructions, before the program control arrives at a matching label for the register $r19$.

Unlike in the previous fault injection instances, where the trigger was placed before round 20, in this case, we place it right before the ARX call for diagonal 4 in round 20 (that is, after the executions for diagonals 1, 2 and 3 are completed). Once the subroutine ARX is invoked, the control spends the initial clock cycles in executing compare and jump instructions before it starts the operations for diagonal 4. It was observed that one of the `brne` instructions got skipped, thereby preventing the control from branching out to the next label. In particular, a `brne` instruction skip was observed following the labels 83, 84 and 85 in Fig. 5.14 (Line numbers 14, 22, and 30). *Targeting line 14 caused the instructions between labels 83 and 44 to be executed twice instead of once*; for correct functionality, *this part of the code should have been skipped on the second occasion*. A similar pattern is detected for the `brne`

```
ARX%=:
cpi r19, 0x00
brne 40f
...
40:                    Column 1 : (0,4,8,12)
cpi r19, 0x00
brne 80f
...
80:
cpi r19,0x01
...
83:
cpi r19, 0x04
brne 44f
...
44:                    Diagonal 1 : (0,5,10,15)
cpi r19, 0x04
brne 84f
...
84:
cpi r19, 0x05
brne 45f
...
45:                    Diagonal 2 : (1,6,11,12)
cpi r19, 0x05
brne 85f
...
85:
cpi r19, 0x06
brne 46f
...
46:                    Diagonal 3 : (2,7,8,13)
cpi r19, 0x06
brne 86f
...
86:
cpi r19, 0x07
brne 47f
...
47:                    Diagonal 4 : (3,4,9,14)
cpi r19, 0x07
brne 87f
...
87:
rjmp ARXExit%=
```

Fig. 5.14 Branching in ChaCha quarterround function

instructions corresponding to the other diagonals too. Recall that the reason for using two jumps per diagonal is because of the limitation on the address range that a `brne` instruction could incorporate. More specifically, a jump from label 83 to 84 would not be possible with a single branch.

The section of the code between labels 83 and 44 perform the *Round* operation on Diagonal 1, except the 7-bit rotation which happens after label 44. The operations that would take place are shown in the equations below.

$$\begin{aligned}
b_0 &= x_0 + x_1, b_3 = (x_3 \oplus b_0) \lll 16 \\
b_2 &= x_2 + b_3, b_1 = (x_1 \oplus b_2) \lll 12 \\
y_0 &= b_0 + b_1, y_3 = (b_3 \oplus y_0) \lll 8 \\
y_2 &= b_2 + y_3, y_1 = (b_1 \oplus y_2)
\end{aligned} \tag{5.13}$$

Further, observe that the inputs (x_0, x_1, x_2, x_3) are nothing but $(s^{20}_{0,0}, s^{20}_{1,1}, s^{20}_{2,2}, s^{20}_{3,3})$, since the fault takes place only the last diagonal undergoes the quarterround operation. Next, we solve the following equations to retrieve the key words k_1 and k_6.

$$\begin{aligned}
x_0 &= z_{0,0} - c_0 \\
x_3 &= z_{3,3} - v_1 \\
y_0 &= z'_{0,0} - c_0 \\
y_3 &= z'_{3,3} - v_1
\end{aligned} \tag{5.14}$$

$$\begin{aligned}
b_3 &= (y_3 \ggg 8) \oplus y_0 \\
b_0 &= (b_3 \ggg 16) \oplus x_3 \\
x_1 &= b_0 - x_0
\end{aligned} \tag{5.15}$$

$$\begin{aligned}
b_1 &= y_0 - b_0 \\
b_2 &= (b_1 \ggg 12) \oplus x_1 \\
x_2 &= b_2 - b_3
\end{aligned} \tag{5.16}$$

$$\begin{aligned}
y_2 &= b_2 + y_3 \\
y_1 &= b_1 \oplus y_2 \\
k_1 &= z'_{1,1} - y_1 = z_{1,1} - x_1 \\
k_6 &= z'_{2,2} - y_2 = z_{2,2} - x_2
\end{aligned} \tag{5.17}$$

As shown in Eq. (5.14), the values of x_0 and x_3 are revealed from the non-faulty keystream. This information is enough to obtain all other x_is, b_is and y_is of the faulty stream, thus reducing the search space of (k_1 and k_6) to a unique pair of values. Following an identical fault injection approach, the values of k_2, k_3, k_4, k_7 can also be retrieved. *Thus the overall key search space is reduced from* 2^{256} to 2^{64} *using three fault injections*. The remaining key words k_0 and k_5 may be further recovered using two additional instruction skips. We have already presented in Sect. 5.6.4 an instruction skip attack on the final addition step in the keystream generation algorithm. Since every quarterround of the keystream generation algorithm also has four 32-bit addition operations, skipping the first and third addition operations in the last quarterround (corresponding to diagonal 4) of round 20 allows us to uniquely retrieve k_0 and k_5 from the following set of equations, respectively:

$$\begin{aligned} b_0'' = x_0, b_3'' &= (x_3 \oplus b_0'') \lll 16 \\ b_2'' = x_2 + b_3'', b_1'' &= (x_1 \oplus b_2'') \lll 12 \\ y_0'' = b_0'' + b_1'', y_3'' &= (b_3'' \oplus y_0'') \lll 8 \\ y_2'' = b_2'' + y_3'', y_1'' &= (b_1'' \oplus y_2'') \lll 7 \end{aligned} \tag{5.18}$$

$$\begin{aligned} b_3'' &= (y_3'' \ggg 8) \oplus y_0'' \\ x_0 &= b_0'' = (b_3'' \ggg 16) \oplus x_3 \\ x_1 &= b_0 - x_0 \\ b_2 &= (b_1 \ggg 12) \oplus x_1 \\ y_2 &= b_2 + y_3 \\ y_1 &= (b_1 \oplus y_2) \lll 7 \end{aligned} \tag{5.19}$$

$$\begin{aligned} b_0 = x_0 + x_1, b_3 &= (x_3 \oplus b_0) \lll 16 \\ b_2 = x_2 + b_3, b_1 &= (x_1 \oplus b_2) \lll 12 \\ y_0' = b_0, y_3' &= (b_3 \oplus y_0') \lll 8 \\ y_2' = b_2 + y_3', y_1' &= (b_1 \oplus y_2') \lll 7 \end{aligned} \tag{5.20}$$

$$\begin{aligned} y_0 &= z_{0,0} - c_0 \\ y_0' &= z_{0,0}' - c_0 \\ y_3 &= z_{3,3} - v_1 \\ b_1 &= y_0 - y_0' \\ b_0 &= y_0 - b_1 \\ b_3 &= (y_3 \ggg 8) \oplus y_0 \\ x_3 &= (b_3 \ggg 16) \oplus b_0 \end{aligned} \tag{5.21}$$

$$\begin{aligned} k_0 &= z_{1,1} - y_1 \\ k_5 &= z_{2,2} - y_2 \end{aligned}$$

By skipping the third addition the values of b_1 and b_0 can be found by solving the Equations in (5.20) and (5.21), and similarly x_0, x_1 can be derived by skipping the first addition, refer to Eqs. (5.18) and (5.19). Every other unknown can be found by substituting these values, including the key words k_0 and k_5. Hence we have uniquely identified the 256-bit key using 5 distinct faults injected targeting different instructions.

5.6.7 An Instruction Replacement Attack on the Final Addition

Contrary to the previous attacks that exploited the instruction skip fault model, our final attack exploits an instruction *replacement* fault, where the intended instruction is replaced by an alternate instruction with the same set of operands. Similar to the attack described in Sect. 5.6.4, this attack also targets the final addition operation between the initial and final states of the keystream generation algorithm. However, instead of skipping word-wise additions, the attack replaces the `add` instruction by a `sub` instruction and `adc` instruction by a `sbc` instruction (the suffix `c` denotes carry-based operations). The attack procedure is described next.

A description of a 32-bit addition module is depicted in Listing 5.3. Observe that while in the first attack, we directly skipped the whole 32-bit addition operation, in this attack, we inject fault at finer granularity on the four 8-bit addition operations within the 32-bit module. Let $s = (s_{31}s_{30}...s_0)$ be one of the words in the final state X^{20} of the keystream generation algorithm, which is added with a corresponding word $k = (k_{31}k_{30}...k_0)$ of the secret key to obtain a word $z = (z_{31}z_{30}...z_0)$ in the output keystream Z. Let $z^1 = (z^1_{31}z^1_{30}...z^1_0)$ be the faulty keystream generated upon skipping the first addition operation, $z^2 = (z^2_{31}z^2_{30}...z^2_0)$ be the faulty keystream generated upon skipping the second addition operation, and so on for z^3 and z^4. The first of these additions typically does not involve a carry as it adds the lowest 8 bits, while the remaining additions involve distinct carry elements.

Listing 5.3 Subroutine *Plus32*

```
Plus32%=:
ld r20, Y
ld r21, Z+
add r20,r21
st Y+, r20
ld r20, Y
ld r21, Z+
adc r20,r21
st Y+, r20
ld r20, Y
ld r21, Z+
```

```
adc r20,r21
st Y+, r20
ld r20, Y
ld r21, Z
adc r20,r21
st Y+, r20
ret
```

The following relations are straightforward to observe upon fault injection in each of the four additions:

$$\begin{aligned} z_7z_6...z_0 &= s_7s_6...s_0 + k_7k_6...k_0 \\ z_{15}z_{14}...z_8 &= s_{15}s_{14}...s_8 + k_{15}k_{14}...k_8 + c_0 \\ z_{23}z_{22}...z_{16} &= s_{23}s_{22}...s_{16} + k_{23}k_{22}...k_{16} + c_1 \\ z_{31}z_{30}...z_{24} &= s_{31}s_{30}...s_{24} + k_{31}k_{30}...k_{24} + c_2 \end{aligned} \tag{5.22}$$

$$\begin{aligned} z^1_7z^1_6...z^1_0 &= s_7s_6...s_0 - k_7k_6...k_0 \\ z^1_{15}z^1_{14}...z^1_8 &= s_{15}s_{14}...s_8 + k_{15}k_{14}...k_8 + c^1_0 \\ z^1_{23}z^1_{22}...z^1_{16} &= s_{23}s_{22}...s_{16} + k_{23}k_{22}...k_{16} + c^1_1 \\ z^1_{31}z^1_{30}...z^1_{24} &= s_{31}s_{30}...s_{24} + k_{31}k_{30}...k_{24} + c^1_2 \end{aligned} \tag{5.23}$$

$$\begin{aligned} z^2_7z^2_6...z^2_0 &= s_7s_6...s_0 + k_7k_6...k_0 \\ z^2_{15}z^2_{14}...z^2_8 &= s_{15}s_{14}...s_8 - k_{15}k_{14}...k_8 + c_0 \\ z^2_{23}z^2_{22}...z^2_{16} &= s_{23}s_{22}...s_{16} + k_{23}k_{22}...k_{16} + c^2_1 \\ z^2_{31}z^2_{30}...z^2_{24} &= s_{31}s_{30}...s_{24} + k_{31}k_{30}...k_{24} + c^2_2 \end{aligned} \tag{5.24}$$

$$\begin{aligned} z^3_7z^3_6...z^3_0 &= s_7s_6...s_0 + k_7k_6...k_0 \\ z^3_{15}z^3_{14}...z^3_8 &= s_{15}s_{14}...s_8 + k_{15}k_{14}...k_8 + c_0 \\ z^3_{23}z^3_{22}...z^3_{16} &= s_{23}s_{22}...s_{16} - k_{23}k_{22}...k_{16} + c_1 \\ z^3_{31}z^3_{30}...z^3_{24} &= s_{31}s_{30}...s_{24} + k_{31}k_{30}...k_{24} + c^3_2 \end{aligned} \tag{5.25}$$

$$\begin{aligned} z^4_7z^4_6...z^4_0 &= s_7s_6...s_0 + k_7k_6...k_0 \\ z^4_{15}z^4_{14}...z^4_8 &= s_{15}s_{14}...s_8 + k_{15}k_{14}...k_8 + c_0 \\ z^4_{23}z^4_{22}...z^4_{16} &= s_{23}s_{22}...s_{16} + k_{23}k_{22}...k_{16} + c_1 \\ z^4_{31}z^4_{30}...z^4_{24} &= s_{31}s_{30}...s_{24} - k_{31}k_{30}...k_{24} + c_2 \end{aligned} \tag{5.26}$$

Solving Eqs. (5.22) and (5.23) together reveals the final state byte $s_7s_6...s_0$, the secret key byte $k_7k_6...k_0$ and the carry c_0. Similarly, solving Eqs. (5.23) and (5.24) would expose the next byte of the final state and the secret key, and so on. The same fault injection method can be followed for the other key words as well, and the whole secret key can thus be recovered with a total of 32 fault injections on an average.

We now present experimental results using the laser-based fault injection setup to validate the aforementioned attacks. Our attack target is an AVR-based implementation of the ChaCha keystream generation algorithm on a 8-bit microcontroller, Atmel ATmega328P, at the decryption site.

5.6.8 Experimental Results: Skipping the Final Addition

This experiment corresponds to the attack described in the Sect. 5.6.4. In this experiment, we set the laser trigger just before final addition commences(that is, right after round 20 ends). The word which gets skipped depends on the timing of the fault injection. The timings for which the 8 key words skip addition are listed in Table 5.5. Let t_0 be the number of clock cycles elapsed from the start of encryption to the beginning of the Final Addition.

5.6.9 Experimental Results: Attack on Rotation

In this experiment we target the 4 relative calls to single-bit circular shift subroutine of 12-bit rotation. Table 5.6 shows four different fault injection timings which would skip one of the four *rcall*s. On an average repeating the experiment twice for each pair of key words would uniquely determine the key pair. Let t_1 be the number of clock cycles elapsed from the start of encryption to the beginning of round 20.

Table 5.5 Fault injection timings for final addition skip

Key word revealed	Fault injection timing (in clock cycles)
k_0	$t_0 + 153$
k_1	$t_0 + 191$
k_2	$t_0 + 230$
k_3	$t_0 + 281$
k_4	$t_0 + 319$
k_5	$t_0 + 359$
k_6	$t_0 + 398$
k_7	$t_0 + 436$

Table 5.6 Fault injection timings for single-bit rotation

Key words revealed	Fault injection timing (in clock cycles)
(k_1, k_6)	$t_1 + 272$
	$t_1 + 301$
	$t_1 + 330$
	$t_1 + 359$
(k_2, k_7)	$t_1 + 1087$
	$t_1 + 1116$
	$t_1 + 1145$
	$t_1 + 1174$
(k_3, k_4)	$t_1 + 1906$
	$t_1 + 1935$
	$t_1 + 1964$
	$t_1 + 1993$
(k_0, k_5)	$t_1 + 2729$
	$t_1 + 2758$
	$t_1 + 2787$
	$t_1 + 2816$

5.6.10 *Experimental Results: Diagonal Fault Attack via Alteration of Control Flow*

This experiment targets branch instructions in specific to be skipped, recall the attack in Sect. 5.6.6. Skipping three `brne` instructions uniquely reveal 6 words of key in total, the details about their timings and the words revealed are mentioned in Table 5.7. Note that t_2 is the time elapsed from the start of the encryption to the beginning of quarterround for the last diagonal ($\sim t_1 + 2445$).

Table 5.7 Fault injection timings for diagonal fault attack

Diagonal repeated	Key words revealed	Fault injection timing (in clock cycles)
$(x_0, x_5, x_{10}, x_{15})$	(k_1,k_6)	$t_2 + 17$
$(x_1, x_6, x_{11}, x_{12})$	(k_2,k_7)	$t_2 + 25$
(x_2, x_7, x_8, x_{13})	(k_3,k_4)	$t_2 + 32$

Table 5.8 Summary of fault attack results on ChaCha

Attack type	Number of fault injections	Key space
Attack on final addition	8	1
Attack on rotation	8	1
Diagonal fault attack	3	2^{64}
	5	1
Instruction replacement	32	1

5.6.11 Experimental Results: Instruction Replacement Attack

In the final experiment, every fault injection leaks 8 bits of the secret key. For instance, to retrieve the key word k_0, we vary the fault injection timings between $t_0 + 153$ and $t_0 + 191$. Changing the timings within this range would cause four different `add` instructions to be skipped and eventually 32 bits of the key is revealed. Similar timing variations reveal other key words as well.

In summary, Table 5.8 shows the number of fault injections needed to obtain the 256-bit key using the aforementioned attacks. Furthermore, a powerful adversary can perform multiple fault injections in the same decryption cycle, to target different key bytes with a single faulty key stream. This would drastically reduce the number of faulty keystreams required to obtain the whole secret key, as all the words are treated independently.

5.7 Summary

In this chapter, we detailed a laser fault injection (LFI) attack technique on a micro-controller. We provided a necessary theoretical background on LFI, together with practical experiments, along with a case study on the ChaCha family of stream ciphers.

There are some parameters which should be taken into consideration in order to increase the chance of success:

- Thickness of a silicon substrate – thickness of a material plays a key role when considering a photon penetration into the silicon.
- Smoothness of a chip surface – if the surface of the back side of a chip is not smooth enough, it can cause refraction and scattering of a laser beam, making a precise attack harder or impossible.
- Beam spot size – the smaller the better. Current manufacturing technologies allow the transistor sizes smaller than 20 nm, therefore it is necessary to have the beam spot size small enough if we want to avoid affecting multiple registers at the same time.

- Precise positioning table – the reason is the same as in the previous case, without a precise positioning it may be impossible to target specific register on a chip.

There are some problems associated with the laser fault injection, which make this attack harder to perform than other fault injection techniques. It is nearly impossible to make a profiling which fits multiple devices, since each chip has a different layout, different manufacturing process and even if we are aiming at the specific chip, de-packaging can cause small but significant differences on the surface which can result in a different fault sensitivity. The second problem is a chip survivability. The microcontroller we used for our experiments was durable enough to withstand several rounds of experiments in a row without any observable damage. This could be due to the old manufacturing process (350 nm), making the connections and transistors large. However, with more advanced technologies, the size of components is much smaller and therefore chips are more vulnerable to optical fault attacks.

Chapter 6
Advanced Fault Attacks in Software: Exploiting the Rowhammer Bug

Sarani Bhattacharya and Debdeep Mukhopadhyay

Abstract In this chapter, we present to the readers a recently reported fault attack technique in the cryptographic literature - attacks exploiting the Rowhammer bug on actual modern-day processors. Rowhammer attacks have exposed a serious vulnerability in modern DRAM chips to induce bit flips in data which is stored in memory. We present here a methodology to combine timing analysis to perform the hammering in a controlled manner to create bit flips in cryptographic keys which are stored in memory. The attack would require only user level privilege for Linux kernel versions before 4.0 and is unaware of the memory location of the key. An intelligent combination of timing *Prime + Probe* attack and *row-buffer collision* is shown to induce bit flip faults in a 1024 bit RSA key on modern processors using realistic number of hammering attempts. This demonstrates the feasibility of fault analysis of ciphers using purely software means on commercial x86 architectures. The attack is also relevant for the newest Linux kernel in a Cross-VM environment where the VMs having root privilege are not denied to access the pagemap.

6.1 What Is Rowhammer?

Rowhammer is a term coined for disturbances observed in recent DRAM devices in which repeated row activation causes the DRAM cells to electrically interact among themselves [78, 98, 101, 177]. This results in bit flips [101] in DRAM due to discharging of the cells in the adjacent rows. DRAM cells are composed of an access transistor and a capacitor which stores charge to represent a bit. Since capacitors loose their charge with time, DRAM cells require to be refreshed within a fixed interval of time referred to as the refresh interval. DRAM comprises of two dimensional array of cells, where each row of cells have its own wordline and for accessing each row, its respective wordline needs to be activated. Whenever some data is requested, the

S. Bhattacharya (✉) · D. Mukhopadhyay
Secured Embedded Architecture Lab, IIT Kharagpur, Kharagpur, India
e-mail: sarani.bhattacharya@cse.iitkgp.ernet.in

S. Patranabis and D. Mukhopadhyay (eds.), *Fault Tolerant Architectures for Cryptography and Hardware Security*, Computer Architecture and Design Methodologies, https://doi.org/10.1007/978-981-10-1387-4_6

cells in the corresponding row are copied to a direct-mapped cache termed Row-buffer. If same row is accessed again, then the contents of row-buffer is read without row activation. However, repeatedly activating a row causes cells in adjacent rows to discharge themselves and results in bit flip.

The authors in [101] show that rowhammer vulnerability exists in majority of the recent commodity DRAM chips and is most prevalent for sub 40 nm memory technology. DRAM process technology over the years have seen continuous reduction in the size of each cell and cell density has also increased to a great extent reducing the cost per bit of memory. Smaller cells hold less charge and this makes them even more vulnerable. Moreover, the increased cell density has a negative impact on DRAM reliability [101] as it introduces electromagnetic coupling effects and leads to such disturbance errors.

6.2 Background Concepts

In this section, we provide a background on some key-concepts, which include some DRAM details, the rowhammer bug and details of cache architecture which have been subjected to attack.

Fig. 6.1 A typical DRAM card

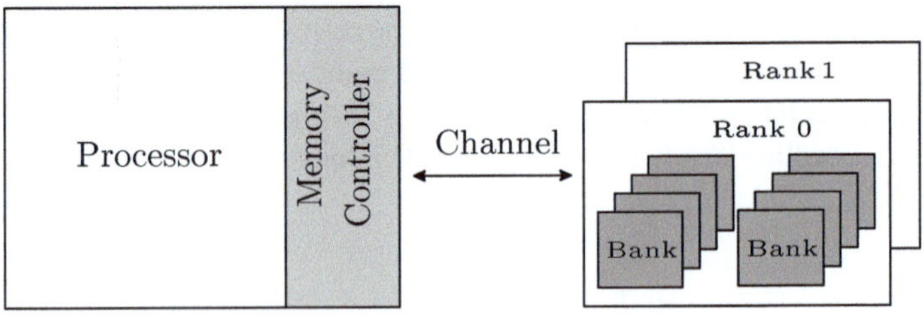

Fig. 6.2 DRAM architecture [101]

6.2.1 Dynamic Random Access Memory

Dynamic Random-Access Memory (DRAM) as in Fig. 6.1 is a Random-Access Memory in which each unit of charge is stored in a capacitor and is associated with an access transistor, together they constitute a *cell* of DRAM. The DRAM cells are organized in rows and columns. The access transistor is connected to the wordline of a row, which when enabled, connects the capacitor to the bitline of the column and allows reading or writing data to the connected row. The reading or writing to cells in a row is done through *row-buffer* which can hold charges for a single row at a time. There are three steps that are performed, when data is requested to be read from a particular row:

Opening Row - The wordline connected to the row is enabled, which allows the capacitors in the entire row to get connected to the bitlines. This results in the charge of each cell to discharge through the bitlines to the row-buffer.
Reading or Writing to cells - The row-buffer data is read or written by the memory controller by accessing respective columns.
Closing Row - The wordline of the respective row is disabled, before some other row is enabled.

6.2.1.1 DRAM Architecture

DRAM is hierarchically composed of Channels, Rank and Banks. The physical link between the memory controller and the DRAM module is termed as *channel.* Inside the channel, the physical memory modules connected to the motherboard are named as *Dual Inline Memory Module (DIMM)* which typically comprises of one or two *ranks* as in Fig. 6.2. Each rank is further comprised of multiple *banks*, as for example, 8 banks exist in a DDR3 rank as in Fig. 6.3. Each bank is a two-dimensional collection of cells having typically 2^{14} to 2^{17} rows and a row-buffer. Any row in a particular bank can only be read and written by involving the row-buffer. The latency in DRAM access when two access request concurrently map to same channel, rank, bank but different row is termed as *row-buffer conflict*. The channel, rank, bank and the row index where a data element is going to reside, is decided as functions of physical address of the concerned data element.

The capacitors in each cell of DRAM discharges with time. The capacitor can hold charge upto a specific interval of time before it completely looses its charge. This interval is termed as *retention time*, which is guaranteed to be 64 ms in DDR3 DRAM specifications [82]. But it is shown, that repeated row activation over a period of time leads to faster discharge of cells in the adjacent rows [101].

In the next subsection, we summarize some key concepts of cache architecture in modern processors.

Fig. 6.3 Single DRAM rank

6.2.2 Cache Memory Architecture

In recent processors, there exists a hierarchy of cache memories where the size of each cache level increases as we move to its higher level, but their access times increases. L3 or Last Level Cache (LLC) is shared across processor cores, takes larger time and is further divided into slices such that it can be accessed by multiple cores concurrently. Figure 6.4 illustrates the architectural specification for a typical Intel Ivy-Bridge architecture [179]. In Intel architecture, the data residing in any of the lower levels of cache are included in the higher levels as well. Thus are inclusive in nature. On a LLC cache miss, the requested element is brought in the cache from the main memory, and the cache miss penalty is much higher compared to the lower level cache hits.

Requested data are brought from the main memory in the cache in chunks of cache lines. This is typically of 64 Bytes in recent processors. The data requested by the processor is associated with a virtual address from the virtual address space allocated to the running process by the Operating System. The virtual address can be partitioned into two parts: the lower bits in Fig. 6.5 is the offset within the page typically represented by $log_2(page\_size)$ bits, while the remaining upper bits forms the *page_number*. The *page_number* forms an index to *page table* and translates to

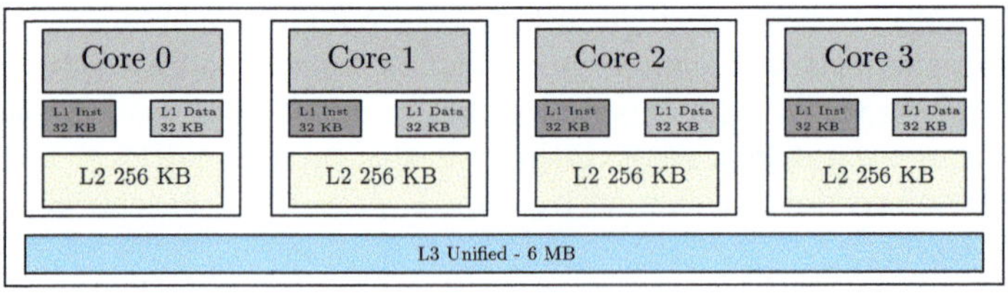

Fig. 6.4 Cache architecture in Intel Ivy Bridge [179]

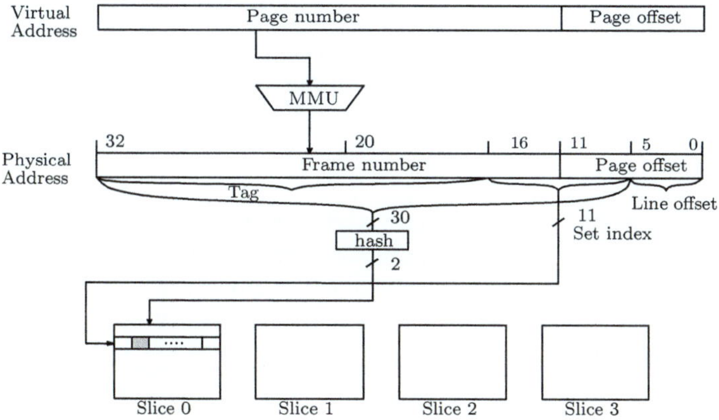

Fig. 6.5 Cache memory indexing [120]

physical *frame number*. The frame number together with the offset bits constitute the physical address of the element. The translation of virtual to physical addresses are performed at run time, thus physical addresses of each elements are most likely to change from one execution to another.

The physical address bits decide the cache sets and slices in which a data is going to reside. If the cache line size is of b bytes, then least significant $log_2(b)$ bits of the physical addresses are used as index within the cache line. If the target system is having k processor cores, the LLC is partitioned into k slices each of the slice have c cache sets where each set is m way associative.

The $log_2(k)$ bits following $log_2(b)$ bits for cache line determines the cache set in which the element is going to reside. Because of associativity, m such cache lines having the identical $log_2(k)$ bits reside in the same set. The slice addressing in modern processors is implemented computing a complex Hash function. Recently, there has been works which reverse engineered [81, 120] the LLC slice addressing function. Reverse engineering on Intel architectures has been attempted in [81] using timing analysis. The functions differ across different architectures and each of these functions are documented via successful reverse engineering in [81, 120].

6.3 Related Work on Rowhammer

It was first demonstrated in [101], that persistent and continuous accesses to DRAM cells, lead the neighboring cells of the accessed cell to electrically interact with each other. The phenomenon of flipping bits in DRAM cells is termed as the *rowhammer* bug [101, 177]. As described in Sect. 6.2.1, accessing a byte in memory involves transferring data from the row into the bank's row-buffer which also involves discharging the row's cells. Repeated discharging and recharging of the cells of a row can result in leakage of charge in the adjacent rows. If this can be repeated enough times, before automatic refreshes of the adjacent rows (which usually occur every 64 ms), this disturbance error in DRAM chip can cause bit flips.

```
Code-hammer
{
  mov (X), %eax   // read from address X
  mov (Y), %ebx   // read from address Y
  clflush (X)     // flush cache for address X
  clflush (Y)     // flush cache for address Y
  jmp Code-hammer
}
```

In [101], a code snippet is provided in which the program generates a read to DRAM on every data access. The *mov* instructions generate access request to DRAM at *X* and *Y* locations. The following `clflush` instructions evict the data elements from all levels of cache. However, if X and Y point to different rows in the same bank, code-hammer will cause X and Y's rows to be repeatedly activated. This may lead to disturbance error in the adjacent cells of the accessed rows, resulting in flipping bits in the adjacent row cells.

In [118], it was first demonstrated that rowhammer not only has issues regarding memory reliability but also are capable of causing serious security breaches. DRAM vulnerability causing bit flips in Native Client (NaCl) program was shown to gain privilege escalation, such that the program control escapes x86 NACL sandbox and acquires the ability to trigger OS syscall. The blog also discusses about how rowhammer can be exploited by user-level programs to gain kernel privileges. In [118], the bit flips are targeted at a page table entry which after flip points to a physical page of the malicious process, thus providing read-write access to cause a kernel privilege escalation. A new approach for triggering rowhammer with x86 non-temporal store instruction is proposed in [148]. The paper discusses an interesting approach by using `libc's memset` and `memcpy` functions and the threats involved in using non-temporal store instructions for triggering rowhammer. A vivid description of the possible attack scenarios exploiting bit flips from rowhammer are also presented in the paper.

A JavaScript based implementation of the rowhammer attack, Rowhammer.js [74] requires a JavaScript enabled website to trigger rowhammer on a remote hardware. This fully automated attack exploits the timing channel in the offline phase to learn the cache address mapping and provides a concrete framework to develop an optimal eviction strategy which proves itself instrumental to flush cache lines from the cache memory. The work also claims to exploit rowhammer bug with high accuracy without using `clflush` instruction. Being implemented in JavaScript, it can induce hardware faults remotely. It exploits the fact that large typed arrays in JavaScript in all recent browsers such as Google Chrome and Firefox are allocated 2 MB pages. It has been mentioned in the paper that, any memory allocation from the scripting languages will be allocated 2 MB pages for large arrays by the operating system. But, in order to mount a successful fault attack on a system the adversary should have the handle to induce fault in locations such that the effect of that fault is useful in making the attack successful.

Another variant of rowhammer exists, termed as double-sided-rowhammer [119]. This variant chooses three adjacent rows at a time, targeting bit blips at the middle row by continuously accessing the outer rows for hammering. The existing implementation [119] is claimed to work in systems having a specific DRAM organization. In [174], the authors show that the traditional rowhammer exploitation techniques do not work on mobile devices and mount a deterministic Rowhammer attack DRAMMER on Android/ARM devices. The exploit uses memory templating technique to probe memory for flippable bits in the memory. In all of the existing works, precisely inducing bit flip in the data used by a co-residing process has not been attempted. None of the previous works attempted to demonstrate a practical fault analysis attack using rowhammer. The address mappings to various components of the LLC and DRAM being functions of physical address bits, inducing bit flip in a data residing in an unknown location of DRAM seems to be a challenging task.

In this chapter, we illustrate a software driven fault attack on public key exponentiation by inducing a bit flip in the secret exponent [29]. It is well known from [35], theoretically if any fault is induced while public key exponentiation is taking place, then a single faulty signature is enough to leak the secret exponent in an unprotected implementation. However, to inflict the fault using rowhammer on the secret exponent to lead to a usable faulty signature requires further investigation. While [118] was able to successfully induce rowhammer flips in the DRAM to cause a fault in a page table entry, the challenge to induce faults to perform a fault attack on a cipher, requires a better understanding of the location of the secret key in the corresponding row of a bank of the DRAM. More recent developments of rowhammering, like double-sided-rowhammer [119], while increasing the probability of a bit flip, cannot be directly applied to the current scenario, as the row where the secret resides can be in any arbitrary location in the target bank. The chance of the memory location for the secret key lying between the rows of the allocated memory for rowhammer is low. In this scenario a double-sided-rowhammer will reduce the probability of a successful exploitable fault. Hence, it is imperative to ascertain the location of the secret exponent before launching the rowhammer. Our novelty is to combine *Prime + Probe* attack and row-buffer collision, detected again through timing channel, to identify the target bank where the secret resides.

We combine knowledge of reverse engineering of LLC slice and DRAM addressing with timing side-channel to determine the bank in which secret resides. We precisely trigger rowhammer to address in the same bank as the secret. This increases probability of bit flip in the secret exponent and the novelty of our work is that we provide series of steps to improve the controllability of fault induction. The overall idea of the attack involves three major steps. The attacker successfully identifies an *eviction set* which is a set of data elements which maps to the same cache set and slice as that of the secret exponent by timing analysis using *Prime + Probe* methodology. This set essentially behaves as an alternative to `clflush` statement of x86 instruction set. The attacker now observes timing measurements to DRAM access to determine the DRAM bank in which the secret resides in. The variation in timing is observed due to the row-buffer conflict forced by the adversary, inducing bit flips by repeated row activation in the particular bank where the secret is residing. Elements which map to same bank but separate rows are accessed continuously to observe a successful bit flip.

In the following sections, we use this knowledge of underlying architecture to build an attack model which can cause successful flips in a secret value.

6.4 Combining Timing Analysis and Rowhammer

In this section we discuss the development of an algorithm to induce bit flip in the secret exponent by combining timing analysis and DRAM vulnerability.

6.4.1 Attack Model

We aim to induce bit fault in the secret exponent of the public key exponentiation algorithm using rowhammer vulnerability of DRAM with increased controllability. The secret resides in some location in the cache memory and also in some unknown location in the main memory. The attacker having user-level privileges in the system, does not have the knowledge of these locations in LLC and DRAM since these location are decided by mapping of physical address bits. The threat model assumed here, allows adversary to send known ciphertext to the algorithm and observe its decrypted output.

Let us assume that the adversary sends input plaintext to the encryption process and observes the output of the encryption. Thus the adversary gets hold of a valid plaintext-ciphertext pair, which will be used while checking for the bit flips. The adversary has the handle to send ciphertext to the decryption oracle, which decrypts the input and sends back the plaintext. The decryption process constantly polls for its input ciphertexts and sends the plaintext to the requesting process. The adversary aims to reproduce bit flip in the exponent and thus first needs to identify the corresponding bank in DRAM in which the secret exponent resides. Let us assume, that the secret exponent resides in some bank say, bank A. Though the decryption process constantly performs exponentiation with accesses to the secret exponent, but such access requests are usually addressed from the cache memory itself since they result in a cache hit. In this scenario it is difficult for the adversary to determine the bank in which the secret resides because the access request from the decryption process hardly results in main memory access.

According to the DRAM architecture, the channel, rank, bank and row addressing of the data elements depend on the physical address of the data elements. In order to perform rowhammering on the secret exponent, precise knowledge of these parameters need to be acquired, which is impossible for an adversary since the adversary does not have the privilege to obtain the corresponding physical addresses to the secret. This motivates the adversary to incorporate a spy process which probes on the execution of the decryption algorithm and uses timing analysis to successfully identify the channel, rank and even the bank where the secret gets mapped to.

The adversary introduces a spy process which runs everytime, before each decryption is requested. The spy process issues accesses to data elements of the eviction set, which eventually flushes the existing cache lines with its own data requests and fills the cache. Thus during the next request to the decryption process, the access to the secret exponent results in a cache miss and the corresponding access request is addressed from the bank A of main memory. Effectively, a spy process running alternate to the decryption process, makes arbitrary data accesses to ensure that every access request from the decryption process is addressed from the corresponding bank of the main memory.

6.4.2 Determining the Eviction Set

As mentioned before, the attack model is assumed to be such that the adversary has access to a system where the decryption is running. The decryption algorithm performs exponentiation involving the secret exponent and initially the adversary aims to determine the cache sets in which the secret exponent bits maps to. The adversary is oblivious of the virtual address space used by the decryption engine and thus involves a spy process which uses *Prime* + *Probe* [113, 139, 179] cache access methodology to identify the target sets. The execution scenario of the decryption and the adversarial processes running concurrently on the same system are depicted in Fig. 6.6. In this context, the spy process targets the Last Level cache (LLC) since it is shared within all cores of the system. The adversary sends input to the decryption engine which performs decryption on request, otherwise remains idle. Figure 6.6 illustrates the following steps.

1. **Step 1**: The adversary starts the spy process, which initially allocates a set of data elements and consults its own pagemap to obtain the corresponding physical addresses for each element. The kernel allows userspace programs to access their own pagemap (`/proc/self/pagemap`)[1] to examine the page tables and related information by reading files in */proc*. The virtual pages are mapped to physical frames and this mapping information is utilized by the spy process in the following steps.
2. **Step 2**: Once the physical addresses are obtained, the Last level cache set number and their corresponding LLC slice mappings are precomputed by the spy with the address mapping functions as explained in Sect. 6.2.2. Let us suppose, that the target system is having k processor cores, thus the LLC is partitioned into k

[1](For all Linux kernels before version 4.0, the versions released from early 2015 requires administrator privilege to consult pagemap, refer to Sect. 6.7.2).

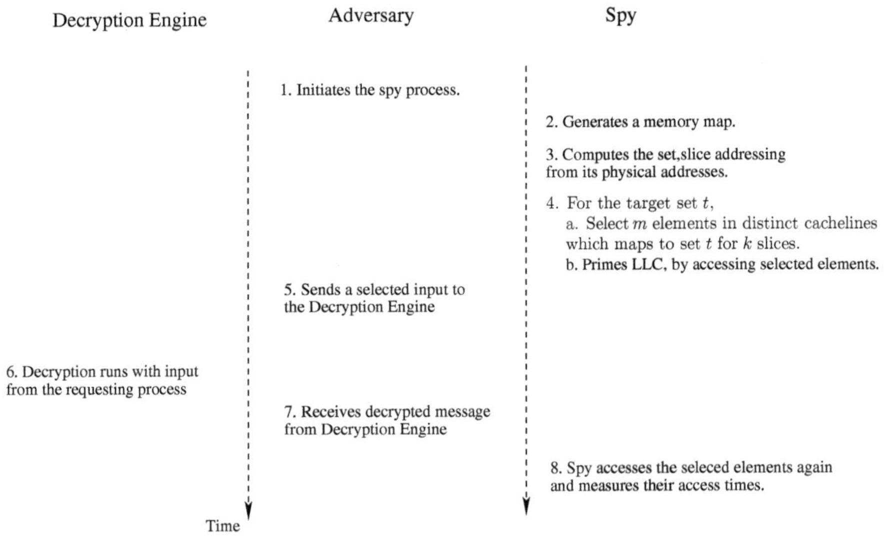

Fig. 6.6 Steps to determine Cache sets shared by secret exponent

slices, each of the slice have c cache sets where each set is m way associative. All elements belonging to the same cache line are fetched at a time.

- If the cache line size is of b bytes, then least significant $log_2(b)$ bits of the physical addresses are used as index within the cache line.
- As described in Sect. 6.2.2, the $log_2(k)$ bits following these $log_2(b)$ bits determine the cache set in which the element is going to reside.
- Because of associativity, m such cache lines having identical $log_2(k)$ bits reside in the same set.
- In modern processors, there exists one more parameter that determines which slice the element belongs to. Computing the Hash function reverse engineered in [81, 120], we can also compute the slice in which a cache set gets mapped. The functions are elaborated in the experimental Sect. 6.5.1.

Thus, at the end of this step, the spy simulates the set number and slice number of each element in its virtual address space.
Repeat the following steps for all of the c sets in the LLC.

3. **Priming Target set** t: The spy primes the target Set t and becomes idle. This is the most crucial step for the entire procedure since the success of correctly determining the cache sets used by spy process entirely depends on how precisely the existing cache lines have been evicted from the cache by the spy in the Prime phase. In order to precisely control the eviction of existing cache lines from set t, a selection algorithm is run by the spy which selects an *eviction set* of $m * k$ elements each belonging to set t from its defined memory map.

- Thus the selection algorithm selects elements belonging to distinct cache lines for each of the k cache slices where their respective physical addresses maps to the same set t. These selected data elements constitutes the *eviction set* for the set t.
- In addition to this, since each set of a slice is m way associative, the selection algorithm selects m such elements corresponding to each k cache slice belonging to set t.
- The spy process accesses each of these $m * k$ selected memory elements repeatedly in order to ensure that the cache replacement policy has evicted all the previously existing cache lines.

This essentially ensures that the target set t of all slices is only occupied with elements accessed by the spy.

4. **Decryption Runs**: The adversary sends the chosen ciphertext for decryption and waits till the decryption engine sends back the message. In this decryption process, some of the cache lines in a particular set where the secret maps, gets evicted to accommodate the cache line of the secret exponent.
5. **Probing LLC**: On getting the decrypted output the adversary signals the spy to start probing and timing measurements are noted. In this probing step, the spy process accesses each of the selected m elements (in Prime phase) of eviction set t for all slices and time to access each of these elements are observed.
The timing measurements will show a variation when the decryption algorithm shares same cache set as the target set t. This is because, after the priming step the adversary allows the decryption process to run. If the cache sets used by the decryption is same as that of the spy, then some of the cache lines previously primed by the spy process gets evicted during the decryption. Thus, when the spy is again allowed to access the same elements, if it takes longer time to access then it is concluded that the cache set has been accessed by the decryption as well. On the other hand, if the cache set has not been used by the decryption, then the time observed during probe phase is less since no elements primed by spy have been evicted in the decryption phase.

6.4.2.1 Determining the LLC Slice Where the Secret Maps

The *Prime* + *Probe* timing analysis elaborated in the previous discussion successfully identifies the LLC set in which the cache line containing the secret exponent resides. Thus at the end of the previous step we obtain an *eviction set* of $m * k$ elements which map to the same set as the secret in all of the k slices. Now, this time the adversary can easily identify the desired LLC slice by iteratively running the same *Prime* + *Probe* protocol separately for each of the k slices with the selected m elements for that particular slice. The timing observations while probing will show significant variation for a set of m elements which corresponds to the same slice where the secret maps. Thus we further refine the size of *eviction set* from $m * k$ to m elements.

6.4.3 Determining the DRAM Bank that the Secret Maps

In this section, we describe a timing side channel analysis performed by the adversary to successfully determine the bank of the DRAM in which the secret exponent maps to. In the previous section, a timing analysis is elaborated which finally returns an *eviction set* of m elements which maps to the same set as well as the same slice as the secret exponent. Thus, if the adversary allows the spy and the decryption engine to run in strict alternation, then the decryption engine will always encounter a cache miss for the secret exponent, and the access request shall always result in a main memory access. As described in Sect. 6.2.1, DRAMs are primarily partitioned into channel, ranks and banks where each bank access is serviced through a buffer termed as row-buffer. Concurrent accesses to different rows in the same DRAM bank results in row-buffer conflict and automatically leads to higher access time. The functions which decide the channel, rank and bank mapping from the physical addresses are not disclosed by the architecture manufacturers. In some recent works, reverse engineering of these unknown mappings have been targeted. A successful deployment of a high speed covert channel has also been reported [143].

We illustrate a timing analysis of accessing separate DRAM banks using this knowledge of reverse engineering and the following steps highlight how this is achieved. In order to exploit timing variation occurring due to the row-buffer collision, accesses requested from the decryption process as well as the adversarial spy process must result in main memory accesses. Intuitively, DRAM access time will increase only if addresses map to the same bank but to a different row. Thus to observe row-buffer conflict between the decryption and adversarial spy the major challenges are:

- To ensure that every access to secret exponent by the decryption process results in LLC cache miss and thus automatically result in main memory access. This is elaborated in the previous subsection, as to how the spy determines the *eviction set* and selectively accesses those elements to evict existing cache lines from the set. Let the spy generates an eviction set C with data elements in distinct cache lines mapping to the same set and slice as the secret.
- This suggests that before each decryption run, the spy has to fill the particular cache set by accessing elements in eviction set C.
- In addition to this, row-buffer conflict and access time delay can only be observed if two independent processes concurrently request data residing in the same bank but in different rows. In order to produce a row-buffer conflict with the secret exponent requested by the spy, the adversary has to produce concurrent access requests to the same bank.

The adversary allows the spy process to `mmap` a chunk of memory and the spy refers to its own pagemap to generate the physical addresses corresponding to each memory element. Following the functions reverse engineered in [79, 143] the spy pre-computes the channel, rank and bank addressing for the corresponding physical addresses. As illustrated in Fig. 6.7, the timing analysis has to be performed by

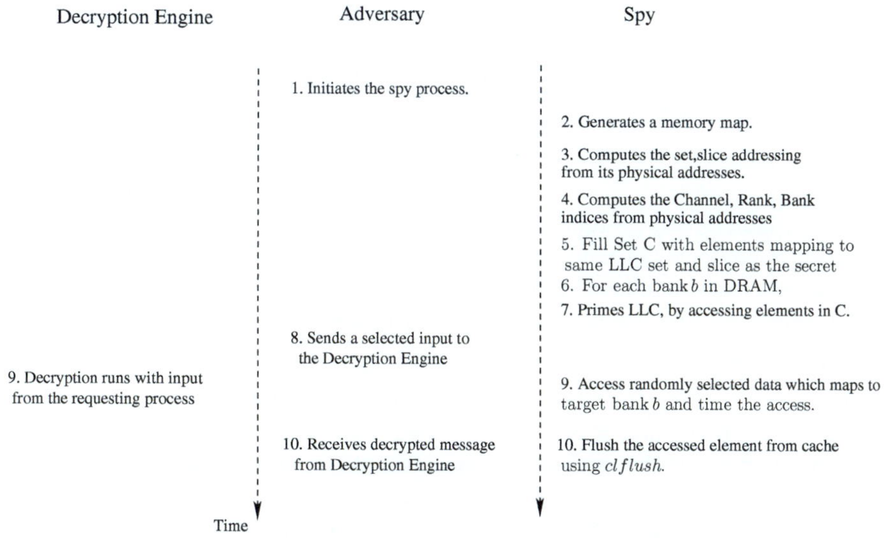

Fig. 6.7 Steps to determine the mapping between the DRAM bank and the secret

accessing elements from each bank. After each access request by the spy, the elements are flushed deliberately from the cache using `clflush`.

The adversary sends an input to the decryption engine and waits for the output to be received. While it waits for the output, the spy process targets one particular bank, selects a data element which maps to the bank and accesses the data element. This triggers concurrent accesses from the spy and the decryption to the DRAM banks. Repeated timing measurements are observed for each of the DRAM bank accesses by the spy, and this process is iterated for elements from each DRAM bank respectively.

6.4.4 Performing Rowhammer in a Controlled Bank

In the previous subsections, we have discussed how the adversary performs timing analysis to determine cache set collision and subsequently use it to determine DRAM bank collisions to identify where the secret data resides. In this section, we aim to induce fault in the secret by repeatedly toggling DRAM rows belonging to the same DRAM bank as the secret.

Inside the DRAM bank, the cells are two-dimensionally aligned in rows and columns. The row index in which any physical address maps is determined by the most significant bits of the physical address. Thus it is absolutely impossible for an adversary to determine the row index of the secret exponent. Thus rowhammer to the secret exponent has to be performed with elements which map to the same DRAM bank as the secret, but on different row indices until and unless the secret exponent is induced with a bit flip.

The original algorithm for rowhammer in [101], can be modified intelligently to achieve this precise bit flip. The algorithm works in following steps:

- A set of addresses are chosen which map to different row but the same bank of DRAM.
- The row indices being a function of the physical address bits are simulated while execution. Elements of random row indices are selected and accessed repeatedly by the adversary to induce bit flips in adjacent rows.
- The detection of bit flip in secret can be done easily, if and only if the output of decryption differs.

The rowhammering attempts required to produce a suitable bit flip on the secret depends on the total number of rows in a bank, since the adversary has no handle to know in which row in the bank the secret exponent is residing. Neither it has handle to place its own `mmap`-ed data deliberately adjacent to the secret, such that it can easily exploit the rowhammer bug. Thus the adversary can only select those elements which belong to the same bank as secret and access them repeatedly to induce bit flips in the secret. To increase the probability of bit flip in the secret exponent, the adversary needs to `mmap` multiple times to generate data which belong to different rows.

6.5 Experimental Validation for Inducing Bit Flips on Secret

In this section we present the validation of our previous discussion through experiments. Our experiments are framed with the following assumptions:

- We target an 1024 bit RSA implementation using square and multiply as the underlying exponentiation algorithm. We have used the standard `GNU-MP` big integer library (version number `2:5.0.2+dfsg-2`) for our implementation. The adversary sends a chosen ciphertext for decryption which involves an exponentiation using the secret exponent.
- The experimentations are performed considering the address bit mappings of Intel Ivy Bridge micro-architecture. These are the line of processors based on the 22 nm manufacturing process. The experiments are performed on Intel Core i5-3470 processor running Ubuntu 12.04 LTS with the kernel version of 3.2.0-79-generic.
- The adversary is assumed to have user-level privileges to the system where decryption process runs. It uses `mmap` to allocate a large chunk of data and accesses its own pagemap (`/proc/self/pagemap`) to get the virtual to physical address mappings. The Linux kernel version for our experimental setup being older than version 4.0, we did not require administrator privileges to perform the entire attack.

6.5.1 Identifying the Cache Set

The experiments being performed on RSA, the 1024 bit exponent resides in consecutive 1024 bit locations in memory. Considering the cache line size as 64 bytes, 1024 bits of secret maps to 2 cache lines. As described in Sect. 6.2.2, 11 bits of physical address from $b_6, b_7, \ldots b_{16}$ refer to the Last Level cache set. Moreover, the papers [81, 120] both talk about reverse engineering of the cache slice selection functions. The authors in paper [81], used *Prime + Probe* methodology to learn the cache slice function, while the authors in [120] monitored events from the performance counters to build the cache slice functions. Though it has been observed that the LLC slice functions reported in these two papers are not same.

We devised a *Prime + Probe* based timing observation setup and wished to identify the target cache set and slice which collides with the secret. Thus we were in the lines of [81] and used the function from [81] in our experiments for *Prime + Probe* based timing observations. As illustrated in the following section, the timing observations using functions from [81] can correctly identify the target cache slice where the secret maps to. Reverse engineering of Last Level Cache (LLC) slice for Intel Ivy Bridge Micro-architecture in [81] uses the following function:

$$b_{17} \oplus b_{18} \oplus b_{20} \oplus b_{22} \oplus b_{24} \oplus b_{25} \oplus b_{26} \oplus b_{27} \oplus b_{28} \oplus b_{30} \oplus b_{32}.$$

But the architectural specification described in [81], documented that their selected system for Ivy Bridge architecture has LLC size of 3 MB. In our experimental setup, instead of 3 MB we had 6 MB LLC with is divided among 4 cores. Thus we adopted the functions documented for Haswell, and the function worked successfully. The functions used for slice selection are:

$$h_0 = b_{17} \oplus b_{18} \oplus b_{20} \oplus b_{22} \oplus b_{24} \oplus b_{25} \oplus b_{26} \oplus b_{27} \oplus b_{28} \oplus b_{30} \oplus b_{32}$$
$$h_1 = b_{18} \oplus b_{19} \oplus b_{21} \oplus b_{23} \oplus b_{25} \oplus b_{27} \oplus b_{29} \oplus b_{30} \oplus b_{31} \oplus b_{32}$$

Our host machine has LLC with 12 way associativity and having 4 cache slices each consisting of 2048 sets. The adversary `mmaps` a large chunk of memory, and consults its own pagemap to obtain the physical addresses corresponding to each element in the memory map. Using the equations mentioned above, the adversarial process simulates the cache set and slice in which their respective physical address points to. The experimental setup being 12-way associative, the selection algorithm for each set and slice selects the *eviction set* with 12 elements belonging to distinct cache lines and mapping to the same set for a particular slice. The host machine having 4 LLC slice for 4 cores selects the eviction set having altogether $12 * 4 = 48$ data elements in distinct cache lines mapped to same set and all 4 slices.

In our experimental setup, the adversary runs the *Prime + Probe* cache access methodology over each of the 2048 sets in each of the LLC slices. Each of the 2048 sets are targeted one after another. The cycle starts with priming a set with elements from eviction set and then allowing the decryption to happen and again observing the timings required for accessing selected elements from the set. The timing observations from the probe phase on 2 such LLC sets are illustrated in Fig. 6.8. The sets are chosen such that one of them is having a collision with the

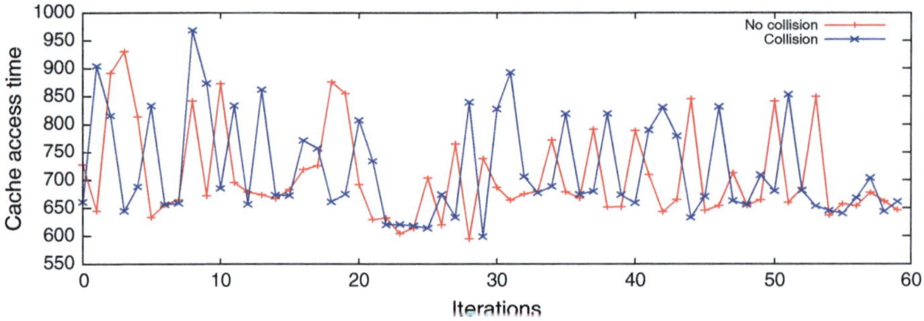

Fig. 6.8 Timing Observations for Cache set collision

secret exponent and the other set does not have any collision. The variation of timing in these two sets is apparent in Fig. 6.8, where the set which observes a collision observes higher access time to the other set. The average access time of the these two sets during the probe phase differs by approximately 80 clock cycles. This implies that the LLC cache set having collision with the secret exponent cache line can be identified from the other sets which does not have collision with the decryption algorithm.

6.5.2 Alternative Strategy to Determine the Target Cache Set

In the previous subsection, we observed that timing observations obtained by repeating the $m * k$ accesses to the individual set on all k slices is sufficient for identification of the target cache set. Though in [74], it has been stated that only $m * k$ accesses may not be sufficient to guarantee the existing cache lines to be evicted from the cache. In this context, we argue that the cache eviction sets are identified in [74] so that accessing the elements of this set in a predetermined order results in an equivalent effect of `clflush` to induce rowhammer flips. Since to exhibit successful bit flips, hammering of rows needs to satisfy various preconditions, it was crucial for authors in [74] to generate an optimal eviction set.

In our case the conditions are little less stringent, since we are using `clflush` to induce bit flips. In addition to this, we constrain the hammering to the target bank. The identification of this bank has a series of experiments to be performed. In this scenario, we claimed that accessing the near-optimal eviction set of $m * k$ accesses for each cache set for all k slices repeatedly will result in eviction of secret from the respective cache set and result in DRAM accesses of the secret key. In addition, we have again performed our experiments by implementing the optimal eviction set as described in [74]. The results we obtain in Fig. 6.10a can be compared with Fig. 6.8. The separation of timing and identification of collision set from the non-collision set definitely improves upon accessing the eviction set with parameters defined in [74].

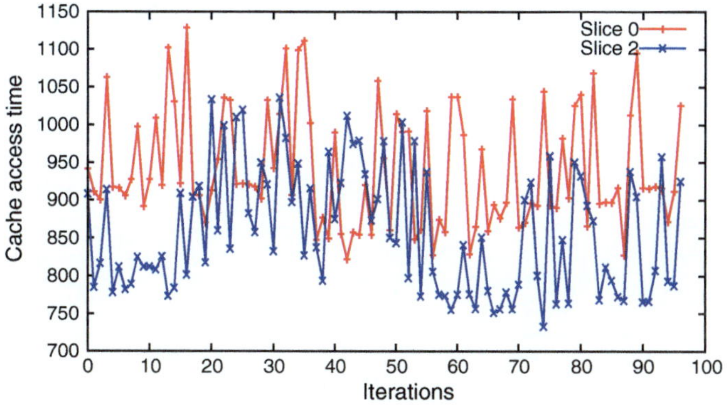

(a) Timing Observations during probe phase when secret maps to slice 0

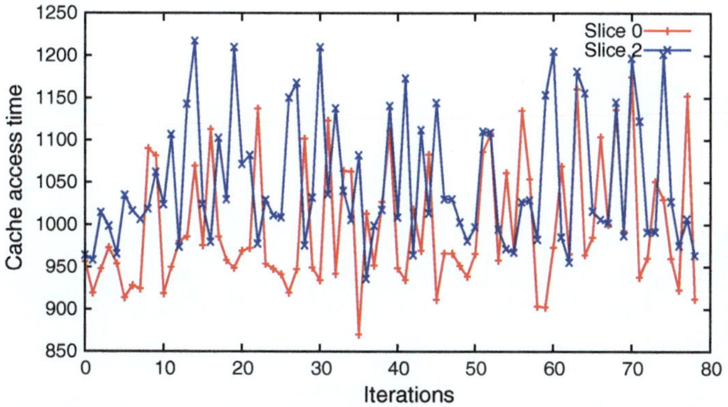

(b) Timing Observations during probe phase when secret maps to slice 2

Fig. 6.9 Timing Observations for LLC slice collision

6.5.3 Identifying the LLC Slice

Once the cache set is identified, the variation from timing observations for different LLC slices leak the information of which LLC slice the secret maps to. In the same experimental setup as in the previous section, we identify the slice in which the actual secret resides, using timing analysis with the slice selection function. Since we have already identified the LLC cache set with which the secret collides, 12 data elements belonging to each slice of the particular set are selected. *Prime* + Probe timing observations are noted for the set of 12 elements for each slice. The slice observing collision with the secret exponent will suffer from cache misses in the probe phase and thus have higher access time to other slices.

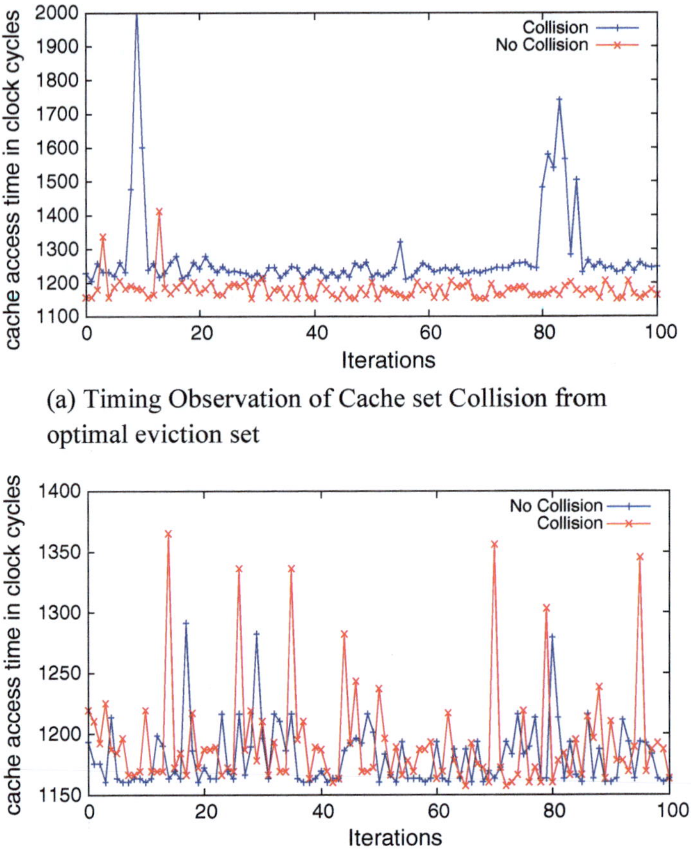

(a) Timing Observation of Cache set Collision from optimal eviction set

(b) Timing Observation of Cache slice Collision using equations in [120]

Fig. 6.10 Timing Observations for LLC set and slice collision

We illustrate the timing observations for two scenarios in Fig. 6.9a, b. In Fig. 6.9a, the secret is mapped to LLC slice 0, while in Fig. 6.9b, the secret gets mapped to LLC slice 2. In both of the figures, access time for probing elements for the cache slice for which the secret access collides is observed to be higher than the other cache slice which belongs to the same set but do not observe cache collision. Thus because of collision of accesses of both the processes to the same slice, the spy observed higher probe time for slice 0, than slice 2 in Fig. 6.9a. On the contrary, in a different run, the secret exponent got mapped to LLC slice 2, which in Fig. 6.9b shows higher probe time than slice 0. Thus we can easily figure out the cache slice for the particular set for which both the decryption and the spy process accesses actually collides.

We also extended our experiment with the reverse engineered cache slice functions from [120]. Figure 6.10b shows the timing observations when we use the slice selection functions for a 4-core processor. The functions [120] are:

$o_0 = b_6 \oplus b_{10} \oplus b_{12} \oplus b_{14} \oplus b_{16} \oplus b_{17} \oplus b_{18} \oplus b_{20} \oplus b_{22} \oplus b_{24} \oplus b_{25} \oplus b_{26} \oplus b_{27} \oplus b_{28} \oplus b_{30} \oplus b_{32} \oplus b_{33}$
$o_1 = b_{07} \oplus b_{11} \oplus b_{13} \oplus b_{15} \oplus b_{17} \oplus b_{19} \oplus b_{20} \oplus b_{21} \oplus b_{22} \oplus b_{23} \oplus b_{24} \oplus b_{26} \oplus b_{28} \oplus b_{29} \oplus b_{31} \oplus b_{33} \oplus b_{34}$

Similar to our previous observations, Fig. 6.10b shows that we were able to identify the target cache slice from the timing observations using cache slice reverse engineering functions from [120].

Determining the LLC set and slice in which secret maps, actually gives the control to the adversary to flush the existing cache lines in these locations, and thus everytime the decryption process have to access the main memory. In simple words, accesses made by the adversary to this particular LLC set and slice acts as an alternative to `clflush` instruction being added to the decryption process.

6.5.4 Identifying the DRAM Bank

From the previous subsections, we identified particularly the LLC set and the slice mappings for the decryption process. Thus if the adversary selects data elements which belong to same set as well slice as to the secret exponent, and alternatively primes the LLC before running each decryption, each time the decryption process will encounter a cache miss and which will eventually get accessed from the main memory. This aids the adversary to identify the respective bank of DRAM, where the secret exponent is mapped. For the 1024-bit RSA exponentiation secret key, the channel, rank and bank mappings of DRAM will be decided by the equations reverse engineered in [79, 143]. In our experimental setup, there exists 2 channel, 1 DIMM per channel, 2 ranks per DIMM, 8 banks per rank and 2^{14} rows per bank.

- The DRAM bank equations for Ivy Bridge [143] is decided by the physical address bits: $ba_0 = b_{14} \oplus b_{18}$, $ba_1 = b_{15} \oplus b_{19}$, $ba_2 = b_{17} \oplus b_{21}$,
- Rank is decided by $r = b_{16} \oplus b_{20}$ and the
- Channel is decided by, $C = b_7 \oplus b_8 \oplus b_9 \oplus b_{12} \oplus b_{13} \oplus b_{18} \oplus b_{19}$.
- The DRAM row index is decided by physical address bits $b_{18}, \ldots, b_{31}$.

In the same experimental setup as previous, the adversary targets each bank at a time and selects elements from the memory map for which the physical addresses map to that particular bank. The following process is repeated to obtain significant timing observations:

1. The spy primes the LLC and requests decryption by sending ciphertext.
2. While the spy wait for the decrypted message, it selects an element for the target bank from the memory map, `clflush`'es it from cache, and accesses the element.

The `clflush` instruction removes the element from all levels of cache, and the following access to the element is addressed from the respective bank of the DRAM.
3. The time to DRAM bank access is also noted.

It is important to note that, there is no explicit synchronization imposed upon the two concurrent processes in their software implementation. The decryption and the spy both requests a DRAM bank access. If the target bank matches with the bank in which the secret is mapped, then we expect to have higher access time. Figure 6.11a, b are the timing observations noted by the spy process while it accesses elements selected from the target bank. Figure 6.11a refers to the case where the higher access times are observed due to the row-buffer collision as the bank accessed by the spy is same as the secret mapped bank. While Fig. 6.11b refers to the situation where the elements accessed by the spy are from an arbitrary different bank than the bank where secret maps. In both of the figures, the significant high peak has been observed in timing range of 350–400 clock cycles. While in Fig. 6.11a, the row-buffer collision is apparent because there are significant number of observations which have timings greater than the region where the peak is observed. Had there been an absolute synchronization of two processes accessing the same DRAM bank, each access to DRAM bank, by either of the two process would have suffered from row-buffer collision. Thus in our scenario, we claim that in majority of cases, though the accesses are addressed from the same bank they seldom result in row-buffer collision, which justifies the peak around 350–400 clock cycles. From this we conclude that detection of row-buffer collision can only be identified over a significant number of timing observations. The DRAM bank which shows such higher access times is identified to be the bank where secret data resides.

6.5.5 Inducing Bit Flip Using Rowhammer

In the previous section, we have illustrated that how the adversary is able to distinguish the bank in which the secret exponent resides. The software implementation of the induction of bit flip is performed by repeated access to the elements of the same bank. The following pseudo-code is used to hammer rows in specific banks After each access to the element it is deliberately flushed from the cache using the `clflush` instruction by the adversary.

```
Code-hammer-specific-bank
{
  Select set of 10 data elements mapping to specific bank
  Repeat
  {
    Access all elements of the set
    Clflush each element of the set
```

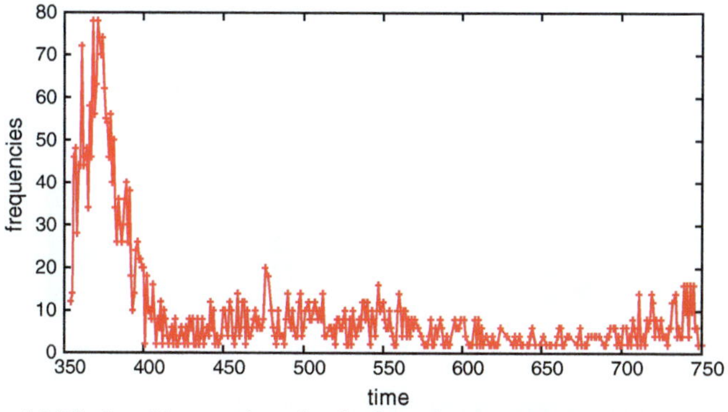

(a) Timing Observations in clock cycles for DRAM bank collision

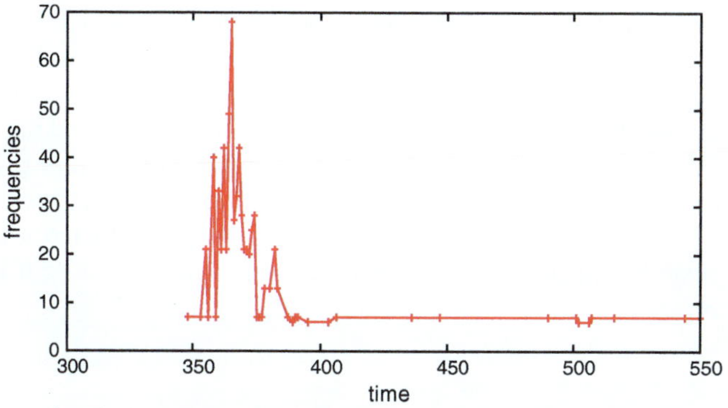

(b) Timing Observations in clock cycles of separate DRAM bank access

Fig. 6.11 Timing Observations for Row-buffer collision during DRAM bank accesses

```
    }
    jmp Code-hammer-specific-bank
}
```

A statistic over observations of bit flips in respective banks is reported in Fig. 6.12. The bar graph shows the number of bit flip instances we were able to observe for respective banks of a single Dual In-line Memory Module (DIMM).

The bit faults that we have observed in our experiments are bit-reset faults.

The row index of the location of the secret in the DRAM bank is determined by the physical address bits of the secret. Thus this implies that the secret exponent can sit in any of the rows in the target bank. Accordingly, we restricted our hammering attempts in the target bank and we selected random accesses to the target bank which eventually resulted in bit flips. Thus we slightly modified our setup such that the code

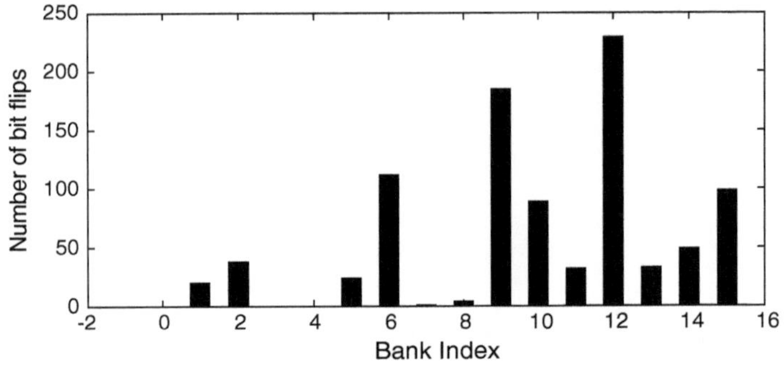

Fig. 6.12 Number of bit flips observed in all banks of a single DIMM

iteratively runs until and unless the decryption output changes, which signifies that secret exponent bits have been successfully flipped. The fault attack in [35] requires a single faulty signature to retrieve the secret. Thus, bit flip introduced in the secret exponent by the rowhammer in a specific bank can successfully reveal the secret by applying fault analysis techniques in [35]. The probability of bit flip is $1/2^{14}$, since there are 2^{14} rows in a particular bank. Interestingly, the size of the secret key has an effect on the probability of bit flip in the secret exponent. In other words, we can say that the probability of bit flip in the secret exponent will be more if the secret exponent size is larger.

6.6 Possible Countermeasures

There has been various countermeasures of rowhammer attacks proposed in literature. In [101], seven potential system level mitigation techniques were proposed which range from designing secure DRAM chips, enforcing ECC protection on them, increasing the refresh rate, identification of victim cells and retiring them and refreshing vulnerable rows (for which the adjacent rows are getting accessed frequently). As mentioned in [101], each of these solutions suffers from the trade-off between feasibility, cost, performance, power and reliability. In particular, the solution named as Probabilistic Adjacent Row Activation (PARA) has the least overhead among the solutions proposed in [101]. The memory controller in PARA is modeled such that every time a row closes, the controller decides to refresh its adjacent rows with probability p (typically $1/2$). Because of its probabilistic nature, the approach is low overhead as it does not require any complex data structure for counting the number of row activations.

Another hardware level mitigation is reported in [118], where it is mentioned that the LPDDR4 standard for the DRAM incorporated two features for the hardware level mitigation such as Targeted Row Refresh (TRR) and Maximum Activate Count (MAC). Among which, it is reported that TRR technique is getting deployed in the next generation DDR4 memory units [83, 121]. TRR incorporates a special module which can track the frequently made row-activations and can selectively refresh the rows adjacent to these aggressor rows. All of the above discussed protections have to be incorporated in hardware, but this does not eliminate the threat from rowhammer attacks since many of the manufacturers refer to these as optional modules.

There are few attempts which provide software level protection from rowhammer attacks. The `clflush` instruction was responsible for removing the target element from the cache and that resulted in DRAM accesses. In order to stop the security breaches from NaCl sandbox escape and privilege escalation [118], Google NaCl sandbox was recently patched to disallow applications from using the `clflush` instruction. The other software level mitigation is to double the refresh rate from 64 to 32 ms by changing the BIOS settings, or alternatively upgrading own BIOS with a patched one. It has been reported in [12], that system manufacturers such as HP, Lenovo and Apple have already considered updating the refresh rate. But both of the techniques such as doubling refresh rate and removing access to `clflush` instruction as a prevention technique has been proved to be ineffective in [12]. The paper illustrates a case study of introducing bit flips inspite of having refresh interval as low as 16 ms and the method does not use the `clflush` instruction. The paper also propose an effective, low cost software based protection mechanism named ANVIL. Instead [12] propose a two-step approach which observe the LLC cache miss statistics from the performance counters for a time interval, and examines if the number of cache misses crosses the predetermined threshold. If there are significantly high number of cache misses, then the second phase of evaluation starts, which samples the DRAM accesses of the suspected process and identifies if rows in a particular DRAM bank is getting accessed. If repeated row activation in same bank is detected, then ANVIL performs a selective refresh on the rows which are vulnerable.

6.7 Further Discussion

The present chapter's main focus is to show that targeted faults can be inflicted by rowhammer. As a consequence, we have cited the example of a fault analysis on RSA, which is not protected by countermeasures. One of the objectives of this chapter, is to show that fault attacks are serious threats even when triggered using software means. This makes the threat more probable as opposed to a fault injection by hardware means: like voltage fluctuations etc. Thus, this emphasizes more the need for countermeasures, at the software level.

Having said that, even standard libraries like OpenSSL use fault countermeasures, but they are not fully protected against these class of attacks. For example, in Black Hat 2012 [26], a hardware based fault injection was shown to be of threat to OpenSSL based RSA signature schemes. It was reported that the initial signature is verified by the public key exponent, however in case of a fault, another signature is generated and this time it is not verified [26]. The final signature is not verified because it is widely assumed that creating a controlled fault on a PC is impractical. More so, the faults are believed to be accidental computational errors, rather than malicious faults. Hence, the probability of inflicting two successive faults is rather low in normal computations! However, in case of rowhammer, as the fault is created in the key, repeating the process would again result in a wrong signature and thus get released.

Hence, the objective of the current article is to highlight that inflicting controlled faults are more probable through software techniques than popularly believed, and hence ensuring that verification should be a compulsory step before releasing signatures.

6.7.1 Assumptions of the Proposed Attack

In our proposed attack, we assumed that the secret decryption exponent resides in a particular location of the DRAM and the decryption oracle continuously polls for input ciphertexts. In addition we also assume that the secret resides in the same location in the DRAM through out the duration of the attack and is not page-swapped by other running processes.

6.7.2 Limitations and Practicality of Our Attack

The access to pagemap is assumed to be available at user privilege level since our setup has 3.2.0-79-generic version of Linux kernel. But from early 2015, for versions of kernel 4.0 onwards, the access to this pagemap has been restricted to processes with root privileges. However, the attack would still be relevant in a cross-VM environment as in [80], where the users of the co-located VMs actually have the administrator privilege and can consult the pagemap for the required virtual to physical address translation. In such a scenario, the attacker is assumed to be mounted on a VM which is co-resident to the VM which hosts the decryption oracle. Timing information obtained from *Prime + Probe* methodology in this experimental setup along with the reverse engineering knowledge can be used to precisely induce fault in the secret of the co-resident VM. We primarily focuses on the vulnerability analysis of rowhammer in context to Linux kernels, but the vulnerability may be equally or more relevant in context to other operating systems where the access to a data structure such as pagemap (in context to Linux kernels) is not restricted only to administrator privileges. Moreover, the attack in its original form might be relevant in customized embedded system applications, thus it would be an interesting exercise to ascertain the security impact of rowhammer in such applications.

6.8 Summary

In this chapter, we illustrated in steps a combination of timing and fault analysis attack exploiting vulnerability of recent DRAM's disturbance error to induce a bit flip targeting the memory bank shared by the secret. This is a practical fault model and uses *Prime* + *Probe* cache access attack methodology to narrow down the search space where the adversary is supposed to induce flip. The experimental results illustrate that the timing analysis shows significant variation and leads to the identification of LLC set and slices. In addition row-buffer collision has been exploited to identify the DRAM bank which holds the secret. The worst case complexity of inducing fault by repeated hammering of rows in the specific memory bank typically is same as the number of rows in bank. The proposed attack finds most relevance in cross-VM setup, where the co-located VMs share the same underlying hardware and thus root privileges are usually granted to the attack instance.

Chapter 7
Automation of Fault Analysis

Sayandeep Saha and Debdeep Mukhopadhyay

Abstract In the previous chapters, the reader has been acquainted with a number of different fault attack methodologies, including differential fault analysis (DFA), fault sensitivity analysis (FSA) and differential fault intensity analysis (DFIA). It is now amply clear that fault analysis attacks typically follow a common strategy of identifying an efficient key distinguisher, resulting from the diffusion of faults, followed by systematic elimination of wrong key candidates. This approach has been adapted for analyzing different ciphers till date, with certain cipher-specific optimizations requiring significant manual effort. An alternative approach, called *Algebraic Fault Analysis* (AFA), which combines so-called algebraic cryptanalysis with standard differential fault analysis, is recently gaining popularity. As a generic automated framework for fault analysis, AFA is found to have a lot of the potential. We dedicate this chapter for the study of AFA in detail, mainly in the context of lightweight block ciphers, for which it has been widely studied.

7.1 Introduction

The advent of the Internet-of-Things (IoT), and Cyber-Physical Systems has lent great impetus to the design of smart devices that are often embedded with electronics, software, sensors, actuators, and network connectivity for easy communication and interaction. While this trend has created a lot of opportunities for the integration of physical world with computing systems, resulting greater economic benefit, it has also mandated the need of putting sensitive information within the devices. This fact leads towards severe security vulnerability issues, which have recently been started to be treated as a design goals for smart systems. Often, cryptography is being chosen as a standard solution for providing provable security against the attacks of different forms. However, resource-hungry nature of cryptographic algorithms has led the

S. Saha (✉) · D. Mukhopadhyay
Secured Embedded Architecture Lab, IIT Kharagpur, Kharagpur, India
e-mail: sahasayandeep@cse.iitkgp.ernet.in

S. Patranabis and D. Mukhopadhyay (eds.), *Fault Tolerant Architectures for Cryptography and Hardware Security*, Computer Architecture and Design Methodologies, https://doi.org/10.1007/978-981-10-1387-4_7

community towards designing lightweight cryptosystems for resource-constrained environments.

Practical attacks e.g. side-channel attacks and fault based cryptanalysis are known to the cryptographic community for quite some time. However, the relevance of them has greatly increased in the recent past, mainly in context of the scenario described in the last paragraph. Several new cryptographic algorithms (e.g. lightweight block ciphers) for resource-constrained environments have been proposed in the last few years. Ensuring security for each of such algorithms are crucial for the sake of their practical usabilities, and that requires each of them, including their implementations to be throughly analyzed against practical attacks. Further, taking the market competitions into consideration, one cannot afford to spend years for such analyses to complete. Unfortunately, with the current status of the cryptographic community, which have mainly kept faith on rigorous mathematical proofs and manual analysis until now, it seems quite an impossible task to be completed within reasonable time.

Taking all such requirements into consideration, complete automation of the cipher analysis techniques against attacks seems to be the only way out. Recently, the fault attack and SCA community has started taking interest in such automation efforts. However, there are several challenges involved in this task, even for the practical attacks, which are quite systematic by nature. The first and foremost among them is the genericness. Ideally an automated tool should work for any given cipher or cryptographic protocol. However, this is extremely difficult to realize in practice for the cryptographic algorithms. Another major challenge is to evaluate the security against all possible class of attacks. In this chapter, we shall examine some of the early efforts towards such automation, mainly in the context of fault attacks on block ciphers.

Recently, some remarkable advances has been made in the fault attack community in the context of automation. The efforts are mainly centered around the so called Algebraic Fault Analysis (AFA), which is found to have enough potential to support this trend in the context of block ciphers. The main idea of AFA stems from Algebraic Cryptanalysis, an alternative form of classical cryptanalysis, popularized in the last decade. In the last six years, AFA has achieved significant success in finding newer attacks on different lightweight ciphers using a generic equation based representation, along with off-the-self solvers like Boolean Satisfiability Solvers (SAT). The goal of this chapter, is to make the readers accustomed to the concept AFA, along with the flavor of automation it provides. For obvious reasons, the discussions will be limited only within the scope of block ciphers. The overall concept of AFA is mathematically intensive. Keeping that in mind, we try to provide some of the necessary mathematical backgrounds required, within the limited scope. After reading the chapter, the readers will have an understanding of the whole concept and would be able to write codes for them on some mathematical software systems like SAGE [62].

The chapter is organized as follows. We begin with an overview of algebraic cryptanalysis in Sect. 7.2. Some important concepts from algebraic cryptanalysis, which are essential for AFA will be presented in Sects. 7.3, 7.4, and 7.5. Eventually, in Sect. 7.6, we shall introduce the concept of AFA in detail along with a summary of the recently published results on it. As the theme of this chapter is fault attack automation, in Sect. 7.7 we shall provide a brief survey of the some other works in this context. Finally, the chapter will be concluded in Sect. 7.8.

7.2 Algebraic Cryptanalysis: An Overview

Algebraic fault analysis (AFA) is an extension of another class of attacks known as algebraic cryptanalysis, which gained a lot of popularity on early 2000's [17, 51, 54, 56]. Before going into the details of AFA, it is thus necessary to provide a brief introduction to this class of attacks, which, by nature, is quite distinct form all other known forms of classical attacks on cryptosystems. A rigorous treatment of this topic is however out-of-the-scope of this chapter and interested readers are referred to [17] for a that. This section will try to provide a rather informal description just by touching the key concepts. Concepts that are relevant in the context of AFA will be presented in subsequent sections in greater detail.

In a very simple sense, algebraic cryptanalysis aims to extract the secret by solving a set of algebraic equations representing a cipher. The motivation for such attacks stems from the seminal paper by Claude E. Shannon on 1949 [165], where it was anticipated that breaking a good cipher should require "as much work as solving a system of simultaneous equations in a large number of unknowns of a complex type", (see [165]). In general, it is not difficult to represent a cipher as a set of nonlinear multivariate algebraic equations, as we will show later in this chapter. However, solving such a system is a nontrivial task, as the computational problem of solving is found to be NP-Complete. As a result, attacks exploiting cipher equations were not considered to be a significant threat in the past – at least until early 90's. The barrier finally started breaking from mid 90's, with successful exploitation of machine solvers like SAT, Gröbner basis etc., which till then were sufficiently matured to handle large problem instances with their sophisticated heuristics. The breakthroughs were first achieved in the field of public key cryptography with the cryptanalysis of Matsumoto-Imai cryptosystem [140] by Patarin and the attack on the basic HFE cryptosystem by Courtois [53]. Gradually people moved their focus towards algebraic cryptanalysis of large block ciphers. However, very recently algebraic cryptanalysis has been applied in the context of public key cryptography, for solving ECDLP problems [145].

As pointed out in the last paragraph, the main challenge in algebraic cryptanalysis lies in solving the cipher equation systems. Typically, block ciphers result in prohibitively large systems of multivariate polynomial equations over finite fields, which are in general found to be overdefined and sparse. Albeit being intractable in general, the computational problem related to solving becomes easier if the system underlying is sufficiently sparse, and this fact is a strong motivating factor for algebraic attacks on block ciphers. However, prominent methods like Gröbner Basis techniques for solving polynomial systems over finite fields fall through, mostly due to the large size of the systems under consideration. First attempts of algebraic cryptanalysis thus primarily focused on developing alternative techniques for handling such systems by somehow exploiting its low degree and sparsity. In EuroCrypt 2000, Courtois et al. [51], presented a general technique called XL for solving such overdefined systems of multivariate polynomial equations over large finite fields. Later in 2002, an improvement of XL, called XSL [56] was proposed by Courtois. Both XL as well as XSL used a very simple key step called linearization. Stated informally, given a non-linear (effectively quadratic; will be explained in later sections) polynomial system, linearization method replaces each monomial having degree >1 with a new variable, making the resulting system completely linear. It can be shown that, the solution set of the original polynomial system is indeed a subset of the solution set of the resulting linear system and linearization destroys no solution of the original system. Original solutions can be found from the linear system's solutions by trial-and-error. However, the success of this entire process critically depends on the number of variables and rank of the resulting linear system. Typically, if the number of variables is almost equal to the rank, the total number of solutions are expected to be small which ease the following trial-and-error step for finding original solutions. To efficiently construct the linear system, XL offered an additional step of lifting the degrees of the system to cubic or higher and adding some extra equations before linearization. Linear systems constructed by XL method are typically found to have small number of solutions making the aforementioned solving process really effective. XSL algorithm does similar thing, in a more sophisticated manner, to properly maintain variable-rank ratio of the linear system near 1.

The consequence of XSL attack was surprising as it was anticipated that, the attack complexity increases linearly with the number of cipher rounds for XSL based attacks in the case of large block ciphers like Rijndeal, Serpent etc., provided the S-Boxs are representable as low degree, sparse polynomial equations. Such a result may have huge impact on traditional block cipher design principles, as it opposes the standard belief built from linear and differential cryptanalysis. In the same paper, Courtois further anticipated without any proof that, it may be possible to attack Rijndeal with roughly a complexity of 2^{100}, which is lesser than the exhaustive search complexity 2^{128}. Although, XSL met with some controversy (including Coppersmith and Moh claiming that such attack cannot work) with its claim of breaking AES, it cultivated the knowledge and in-depth understanding of the block cipher polynomial systems and described circumstances where they might be easy to solve. Further, improvements to XL/XSL were also proposed, leading to MutantXL [42], ElimLin [54] etc. However, with the huge success of SAT solvers, the community gradually moved towards using them. We discontinue the study of these customized solvers here and adapt SAT solvers in the rest of this chapter.

Although, solvers play an important role in algebraic cryptanalysis they are not the only important thing. Proper representation of the cipher equations also play a very crucial role. The most tricky part of this is the representation of S-Boxes. It has been found that, a so-called improper representation of the S-Boxes may even result to the failure of the attack. Making it worse, knowing a representation to be the best for attack is an open problem and often realized by means of trial-and-error. In their paper [54], Courtois suggested some efficient methods for representing DES

S-Boxes and figured out some representations best suited for their purpose. Similar things were done for Rijndeal and Serpent in [56]. However, finding the best S-Box representation for any given cipher is still considered to be an art! In the subsequent sections, we will present a general strategy for creating S-Box equations given their tabular representation. Note that, we do not claim S-Box equations generated by this strategy to be the best. In fact, it can be shown that the best representation for the Rijndeal S-Box cannot by generated using this strategy.

In the following sections, we will describe some of the key concepts from algebraic cryptanalysis in detail. Section. 7.3 will be dedicated towards a formal treatment of the system of polynomial equations over finite fields. We will typically denote such polynomial systems as MC or MQ (when the system is quadratic). In Sect. 7.4, techniques for generating S-Box polynomials will be provided. Finally, in Sect. 7.5, we will present algorithms to convert MQ systems in Algebraic Normal Form (ANF) to Conjunctive Normal Form (CNF).

7.3 Polynomial Systems of Equations over Finite Fields

Systems of polynomial equations over finite fields is a topic having numerous applications in pure mathematics, cryptography, communication and computer science. The subject has been extensively studied by mathematicians in various forms, almost from 3rd century AD. Rich theories and tools exist for studying polynomial systems, especially in the field of Algebraic Geometry (Gröbner Basis, Ideals and Varieties are directly involved with polynomial systems). Gröbner Basis solvers are also utilized in the algebraic cryptanalysis community, mainly due to their theoretical soundness [57]. However, their utility is found to be limited, as they cannot handle very large systems. Interested readers are referred to the book *Ideals, Varieties, and Algorithms* by Cox, Little and O'Shea [57] for a rigorous treatment of Gröbner Basis.

The main goal of this section is to study multivariate polynomial systems over finite fields – more specifically, on $\mathbb{GF}(2)$ and its extensions. It is important to note that, while talking about solutions of the such systems, one must specify the underlying field on which the solutions are expected. For example, given a system of polynomial equations on $\mathbb{GF}(2^n)$, one may be interested *only in the solutions lying on* $\mathbb{GF}(2^n)$. But if he/she is interested in the solutions of the same system on $\mathbb{GF}(2^m)$ with $n|m$, it must be explicitly specified. Throughout with book, we try to follow this convention.

Polynomials over $\mathbb{GF}(2)$ are of particular interest in all variants of algebraic attacks. Mathematically, every n variable polynomial on $\mathbb{GF}(2)$ is a member of a polynomial Ideal of the following form:

$$\mathbb{GF}(2)[x_1, x_2, \ldots x_n]/(x_1^2 - x_1, x_2^2 - x_2, \ldots, x_n^2 - x_n)$$

The representation clearly implies that no square, cubic or even higher degree terms on a single variable can appear for those polynomials. For the sake of completeness, let us mention that, the a polynomial over $\mathbb{GF}(2)$ is XOR sums of monomials containing at most d terms, where d is the degree of the polynomial. The coefficients are either 0 or 1 and one need not to explicitly specify them. As an example, consider the polynomial in 3 variables x_1, x_2, x_3:

$$f : x_1x_2 + x_1 + x_3 + 1$$

The polynomial contains 4 monomials: one of degree 2, two of degree 1 and one of degree 0. Further notice that, the coefficient of the 0 degree monomial is 1.

Throughout this chapter, we deal with polynomial equations. In the context of $\mathbb{GF}(2)$ polynomials, equations can be constructed by equating them to 0 or 1. For example, one may write:

$$x_1x_2 + x_1 + x_3 + 1 = 0$$

or

$$x_1x_2 + x_1 + x_3 + 1 = 1$$

However, for the second equation, note that the 1 in the RHS can be brought to the LHS, canceling the other 1, which was there in the LHS from the beginning. The resulting equation is the following:

$$x_1x_2 + x_1 + x_3 = 0$$

It is a standard practice in the algebraic attack community to represent polynomial equations in the "equal-to 0" form. Also, in most of the cases the "$= 0$" part is omitted while representing equations. Readers should not get confused with this standard practice as we are also going to use it throughout this chapter.

7.3.1 Degree Reduction Algorithm for Polynomials

When solving polynomial systems of equations, the number of equations, number of variables, as well as the maximum degree are some of the crucial measures of difficulty. We shall discuss the first two later in this section. In this subsection, we show how to deal with the high degree equations in a polynomial system. Fortunately, it can be proved that, given a system of polynomial equations (note that, this statement is not limited to finite fields and holds well for polynomials on other algebraic structures as well.), there always exist another second system of polynomials, with a solution set in bijection with the first, so that the degree of all equations in the second system is at most two. Further, the number of new variables and equations added to the original system to accomplish this is polynomial in the number of original variables, for any fixed maximum degree of the original system.

The basic algorithm to accomplish this is very simple and intuitive. We explain this with the help of an example. Consider the following equation in 5 variables:

$$f : d + e + ac + ae + bc + be + cd + de + ade + ace + abd + abc$$

The equation has a maximum degree of 3. However, one can readily observe that it can be reduced to 2 by adding two extra variables α and β as shown below.

$$\begin{aligned}
& d + e + ac + ae + bc + be + cd + de + d\beta + c\beta + \alpha d + \alpha c \\
& \alpha = ab \\
& \beta = ae
\end{aligned}$$

Please notice that we have intentionally dropped the "= 0" terms here following the standard practice. This simple example is adapted from [17], which also provides a neat algorithm for degree dropping, along with its complexity evaluation. The key idea of the algorithm is to replace the monomials having degree $d \geq 3$ in the system with quadratic monomials. Let us consider one such monomial $w = x_{a_1}x_{a_2} \ldots x_{a_d}$, where $a_1, a_2, \ldots a_d$ are indices in the range $[1, n]$. The algorithm introduces $d - 2$ new variables $r_1, r_2, \ldots r_{d-2}$ as dummy variables along with the following equations:

$$\begin{aligned}
r_1 &= x_{a_1}x_{a_2} \\
r_2 &= r_1x_{a_3} \\
r_3 &= r_2x_{a_4} \\
\vdots & \quad \vdots \quad \vdots \\
r_{d-3} &= r_{d-4}x_{a_{d-2}} \\
r_{d-2} &= r_{d-3}x_{a_{d-1}} \\
w &= r_{d-2}x_{a_d}
\end{aligned}$$

Each monomial in the system under consideration has to be replaced following the strategy presented above. The total number of new variables and equations added in the resulting system can also be estimated precisely. We present all the key results for this algorithm the form of the following theorem without stating the proof.

Theorem 7.1 *Let* $f_1(x_1, x_2, \ldots, x_n) = 0,\ f_2(x_1, x_2, \ldots, x_n) = 0,\ \ldots,\ f_m(x_1, x_2, \ldots, x_n) = 0$ *be a system of polynomials having n variables and m equations with maximum degree* $d \geq 3$. *Then there exists another system of equations* $g_1(x_1, x_2, \ldots, x_{n'}) = 0,\ g_2(x_1, x_2, \ldots, x_{n'}) = 0,\ \ldots,\ g_{m'}(x_1, x_2, \ldots, x_{n'}) = 0$ *with degree at most 2, whose solutions are in bijection with those of the system of the f's. Further, let z be the number of monomials of degree 3 or higher in the system of f's, with non-zero coefficients, and let the sum of degrees of all such monomials be D. Then* $m' \leq m + D - 2z$ *and* $n' \leq n + D - 2z$.

Unfortunately, the complexity of this simple degree dropping algorithm for any n variable system is $\Theta(e^{n/2}n^{n/2})$ [17], if the maximum degree Δ of the system is not fixed. However, fixing Δ to some predefined value makes the runtime $\Theta(n^{\Delta})$, which is indeed polytime. It is worth to mention that, there exists a greedy variant of this algorithm, which provides further performance improvements by reducing the number of dummy variables by making careful choice the monomials for replacement. Interested readers are requested to refer [17] for further details.

7.3.2 NP-Completeness of Polynomial System Solving

This subsection is of great theoretical interest from computational point of view. Unlike the previous subsection and the subsequent subsection, it does not provide any algorithmic tool for managing polynomial systems as such. However, the result developed in this section motivates one to adapt SAT solvers for polynomial system solving, which, so far has been proven to be the most effective tool in this domain. Similar to the previous section, we will only state the key results here without proving them explicitly. However, some informal proof sketch will be provided.

The problem of solving a polynomial system of equations over a finite field or over rationals is NP-Complete. As shown in the last subsection, one can easily convert a system having a maximum degree $d \geq 3$ to a system having maximum degree 2. We write the computational problem of solving a system of maximum degree 2 as MQ. Similarly, for any general system with maximum degree at least 2, we write it as MP and for a cubic system, it is written as MC. Further, the associated decision problems are named as MQD, MPD and MCD, respectively. Next, we state the main result of this section in the form of a theorem as follows:

Theorem 7.2 *Deciding whether any cubic polynomial system (MC) over* $\mathbb{GF}(2)$ *has a solution in the base-field is NP-Complete.*

Before providing an informal sketch of the proof of this theorem, let us first look at the consequences of it. Proving the MC decision problem over $\mathbb{GF}(2)$ to be NP-Complete directly proves the same for MP over $\mathbb{GF}(2)$, as any MP problem over $\mathbb{GF}(2)$ can be converted to an equivalent MC problem over $\mathbb{GF}(2)$ using the degree-dropping algorithms. The same claim is true for MQs as well. To see this, assume that MQ over $\mathbb{GF}(2)$ is solvable in polynomial time. As any MC or MP instance is polytime convertable to some equivalent MQ instance, polytime solvability of MQ directly implies polytime solvability of MC and MP, contradicting the claim of the theorem. Finally, the result can be extended for any finite field $\mathbb{GF}(q)$, other than $\mathbb{GF}(2)$. The proof idea is to view any polynomial system over $\mathbb{GF}(q)$ as a mapping from $\mathbb{GF}(q)$ to $\{0, 1\}$. It can be proved that all such mappings from one finite set to another can be represented by a system of polynomials over $\mathbb{GF}(p)$ for any prime p. Taking $p = q$ will then serve our purpose.

Now, let us give a proof sketch for Theorem 7.2. Like any other NP-Completeness proof, we start with an assumption that let there exists some polytime algorithm, which solves any instance of the MC decision problem. Now, we shall use this polytime algorithm to solve a 3-CNF SAT problem instance, which is known to be NP-Complete. This will lead to a contradiction, ultimately proving Theorem 7.2. The only thing remaining to be shown is a polytime conversion strategy from any 3-CNF SAT instance to some MC instance. Let us explain this step by the following example adapted from [17]. We start with a 3-CNF clause $(a \vee b \vee c)$ and eventually convert it as shown below.

$$\begin{aligned}
(a \vee b \vee c) &\Leftrightarrow ((a \vee b) \wedge c) \oplus (a \vee b) \oplus c \\
&\Leftrightarrow ((a \wedge b) \vee (a \wedge c)) \oplus (a \vee b) \oplus c \\
&\Leftrightarrow ((a \wedge c \wedge b \wedge c) \oplus (a \wedge c) \oplus (b \wedge c)) \oplus ((a \wedge b) \oplus a \oplus b) \oplus c \\
&\Leftrightarrow (a \wedge b \wedge c) \oplus (a \wedge c) \oplus (b \wedge c) \oplus (a \wedge b) \oplus a \oplus b \oplus c \\
&\Leftrightarrow (abc + ac + bc + ab + a + b + c) = 1 \\
&\Leftrightarrow \ abc + ac + bc + ab + a + b + c + 1 = 0
\end{aligned}$$

If a is negated in the 3-CNF, it can be replaced by $1 + a$ in the polynomial representation. Hence, it follows that, each 3-CNF clause can be represented as a polynomial equation. This representation of polynomials is often termed as *Algebraic Normal Form* (ANF). In this chapter, we are going to use this term extensively.

7.3.3 Role of Sparsity in MQ Solving

The final phase of this section presents a very important measure to understand the difficulty of polynomial system solving. We have already pointed this out at the beginning of this section. A polynomial system (without loss of generality, one can only talk about MQ systems) can be characterized by two parameters, namely the number of variables (n) and the number of equations (m) in it. We are particularly interested in how these two parameters regulate the difficulty of the solving. Being an NP-Complete problem, however, no exact measure can be given. Still it was found that these two parameters have some roles in deciding the difficulty. As a historical note, in [51, 102] Shamir showed that even if the MQ problem is NP-Hard, its complexity reduces significantly when it is overdefined. Similar claim was made by Courtois [56], in the context of sparsity.

Let us consider the ratio $\gamma = m/n$. We also denote the sparsity of the system by the term β. The system is called *overdetermined* if $\gamma > 1$. Typically, if $\gamma \approx n$ (that is to say $m \approx n^2$), the system is found to be easier to solve in general, by most of the solvers including the SAT-solvers. However, the complete picture is still not exposed. It is also found that the sparsity β has significant roles to play, while the difficulty of solving is concerned. Defining the sparsity is however, not an easy task. Typically, only may think of it as the expected probability of the occurrence of a monomial in the system. Let us discuss this in greater detail. Given an MQ system of m equations and n variables, it is called *ultra-sparse* if it contains only

$\alpha log(n)$ number of unique, degree-2 monomials, where α is a constant. Note that, the number of times such monomials occur is not of concern at this point. Intuitively, if one chooses to replace these monomials with new dummy variables, converting the system to a linear one (this is called linearization as described in Sect. 7.2), such variables will assume total $2^{\alpha log(2)} = n^{\alpha}$ possible values. Now, for a particular choice of values for these quadratic monomials, one may choose to solve the remaining linear system by simple Gaussian-Elimination, having complexity $\Theta(mn\ min(m, n))$ for a system of m equations and n variables. Combining the guessing step and the following Gaussian-Elimination step, the total running time becomes $\Theta(mn^{\alpha+1}\ min(m, n))$, which is indeed polynomial in m and n. This apparently simple strategy explains the role of sparsity very well in MQ system solving.

Systems containing only $\alpha log(n)$ monomials are indeed reasonable. However, we have still not considered the multiple occurrences of these monomials in the system. Let us now consider the number of occurrences of each unique degree-2 monomial in the system. Clearly, a monomial cannot appear more than once in a single equation and we need to consider occurrences across equations only. If the system has n variables then there are $(n^2 + n)/2$ possible quadratic and linear monomials in total plus one constant monomial. If we assume that, in particular, $3/4$ of the unique $\alpha log(n)$ monomials will appear randomly in each equation, the sparsity of the system turns out to be:

$$\beta = \frac{4n + 3\alpha\ log(n)}{2n^2 + 2n + 1} \approx \frac{1}{2n} \tag{7.1}$$

Here the sparsity is calculated simply by taking the ratio of total occurring monomials in the system to the total number of possible monomials, which is quite low in value here in this case. As it appears, the sparsity can be used as a metric for measuring the difficulty of solving. This simple study was performed by G. Bard in his nice book [17].

Although, it is often anticipated that the sparsity is a strong representative of system solvability, the relationship is still not very clear. One simple analogy with the linear systems can be drawn, where it is found that the sparse linear systems are typically easier to solve than the dense systems of same size. Since, many of the solving tools including XL, Gröbner Basis use linear algebra, sparse systems are expected to be easier for them to handle. Further, considering the Gröbner Basis algorithms, the limiting factor in solving is the memory. The memory efficiency can be improved significantly if one deals with a sparse system rather than a dense system. Finally, it is found that SAT solvers work remarkably well with sparse systems than the dense ones. The reason is however not very well understood.

7.4 Generation of S-Box Polynomials

So far in this chapter, we have discussed some basic properties and tools related to polynomials over finite fields. Until now, no methods were provided which enable one to write equation systems for state-of-the-art block ciphers. Note that, by cipher equations, we only mean polynomial systems over $\mathbb{GF}(2)$ here. Although, compact equations over extension fields can be written for some block ciphers (the best example is AES; please see [47]), presently we have no efficient means of solving such systems. Strictly speaking, there is no best way to write equations for block ciphers. As we have already mentioned in Sect. 7.2 that, deciding a particular equation representation to be the best is an open problem. However, some hope still exist and there are some standard practice for this art, which are worth to mention. There is no doubt of this fact that, writing the S-Boxes are the trickiest part of cipher equation construction. For example, given one known plaintext, the complete AES equation system can be written as a system of about 4000 multivariate quadratic (MQ) equations. It was shown by Courtois et al. in [56], that AES inversion operations are representable with a very compact set of degree two equations. Although, AES was shown to be unbreakable practically, there is a body of positive results for block, stream and public key ciphers which were broken with such algebraic attacks – mostly due to the fact that their S-Box representations were sparse and low degree. Courtois have further shown in [54], that the DES S-Boxes, which are otherwise considered very "well-built" in the context of differential cryptanalysis, are vulnerable under algebraic attacks if reduced round (up to 6) versions of DES are considered. Later in this chapter, we shall see the impact of this result in context of the AFA attacks on DES.

Courtois et al. proposed several representations of the DES S-Boxes in [54], useful for algebraic attacks. The representations were broadly classified into 3 classes. Class I deals with low degree multivariate I/O equations. I/O degree is defined as the smallest degree of the relations of form $g(x_0, x_1, \ldots, x_{n-1}; y_0, y_1, \ldots, y_{m-1}) = 0$, where, $(x_0, x_1, \ldots, x_{n-1})$ and $(y_0, y_1, \ldots, y_{m-1})$ are the input and output variables of some $n \times m$ S-Box. Such relations qualify as I/O relations only if they hold with probability 1 for their corresponding S-Boxes. Courtois further, pointed out that for any $n \times m$ S-Box, such I/O relations exist, and can be counted and obtained in a very efficient manner. We do not provide the results here but they can be found in [54]. Although, sufficient number of linearly independent, fully cubic or degree-4 equations can be obtained through this method, they were found being very dense, having large number of equations. As a remedy, another class (Class-II) of relatively sparse equations, with probably higher degree were proposed. Class-II equations are constructed in a way so that their maximum degree remains ≤ 4, while the number of monomials per equation also remains typically within 4–5. Further, the linear independence of equations are not guaranteed in these cases. Given all these constraints, approximately 500–700 degree 4 equations were obtained for each DES-S-Box. Finally, in the Class-III, probably the fittest equations were obtained. Here one starts with standard cubic representations of any DES S-Box and adds extra variables to reduce the degree while increasing the sparsity. It was found that for DES, each

S-Box requires about 52 variables for a highly sparse representation. Further, the degree becomes quadratic and each equation contains effectively 0 or 1 nonlinear monomial. This representation is in-sync with the smallest hardware representation of DES S-Boxes.

The idea of adding extra variables to reduce degree and increase sparsity seems to be a nice trick for constructing good equations. However, we still lack a general way of constructing S-Box equations. In this paragraph, we present one such technique, which is due to Knudsen and Miolane [103]. and has gained much popularity in the AFA community. The method finds out equations for a $n \times m$ S-Box having maximum degree d, decided a-priori, where n and m are measured in bits. The idea is to construct a binary matrix A with 2^n rows and $\sum_{i=0}^{d} \binom{n+m}{i}$ columns. The 2^n inputs of the S-Box are the row entries and the $\sum_{i=0}^{d} \binom{n+m}{i}$ ordered monomials (according to some graded monomial ordering; please see [57] for monomial orderings in Gröbner Basis), of degree less than or equal to d are the column entries. An entry a_{ij} in A is set if for the input i of the S-Box, the monomial at column j assumes value 1. All deterministic S-Box equations of degree less than or equal to d, are then obtained by calculating the basis of the null space of A. A total of $\sum_{i=0}^{d} \binom{n+m}{i} - 2^n$ are guaranteed by this process.

The process described in the last paragraph can be further enhanced by the calculation of the Gröbner Basis of the obtained equations. It is often found to be an useful practice. It is worth to mention that, the quality of the equations, obtained from the above mentioned method, crucially depends on the chosen monomial ordering. Usually, it is found that *graded reverse lexicographic ordering* provides the most compact representation. However, it is not always the case. One can still figure out better S-Box representations computed manually. For example, the most compact S-Box representation for AES can never be obtained by this technique. Same is true for DES as well. The quadratic equations for AES S-Boxes are obtained by considering the properties of finite field inversion, whereas for DES, they were specially crafted, as we already have described. Gröbner Basis and the monomial orderings play an important role in S-Box representations as mentioned in the beginning of this paragraph. However, a rigorous discussion of this topic is out-of-the-scope. We strongly recommend the readers to go through the first 3 chapters of [57], to get a proper understanding of this topic.

While we have precisely described how to construct S-Box equations, nothing has been told so far, for the linear equations of the system. We find them usually very easy to describe. It will become clear, once we describe a complete equation system for a cipher. However, there are a few technicalities involved while handling ciphers described over finite fields (like AES or LED). In such cases, the linear layers are usually described in word level, and most of the cases they involve an MDS matrix. Writing $\mathbb{GF}(2)$ equations for such systems is a bit tricky, as one needs to convert all diffusion layer operations, effectively into bit matrices represented in $\mathbb{GF}(2)$. Fortunately, there are standard ways to do such conversions, at least for ciphers represented over relatively small extension fields, like AES. The monograph by Carlos Cid [47], systematically describes the procedure for AES. For a general methodology, please refer to Chap. 15 of [17].

7.5 MQ to CNF Conversion

There are several ways available for solving multivariate polynomial systems. In fact, the initial research goal of algebraic cryptanalysis was to engineer best possible solvers for this purpose. However, soon people started moving towards SAT solvers, not only due to their efficiency; but also for the ease of use. Modern SAT solvers are probably the most widely used tools in the computer science and applied mathematics community. The continuous improvements in the search heuristics of SAT solvers make them extremely powerful and robust. Standard SAT solvers, nowadays can easily handle millions of variables and clauses. But as a cost of this extreme efficiency, it is really very hard to penetrate inside the subject. Although, understanding the working principles of modern SAT solvers require special expertise, using them as black box is extremely easy. For our purpose, we only need them as black boxes. So, in this chapter, we intentionally skip the working strategies of SAT solvers. Interested readers may refer to the existing literature. Also, the blog of Mate Soos, the author of the famous CryptoMiniSAT solver [168], provides some idea on the working principles of modern SAT solvers.

Although, one can safely skip the details of the SAT solvers, we need to know how to feed the cipher equation systems to them. Typically, SAT solvers take inputs in *Conjunctive Normal Form* (CNF). A CNF formula is a conjunction (AND) of *clauses*, which are disjunction of positive or negated variables called *literals*. Converting cipher equations, which are in ANF form, to CNF requires significant algorithmic effort, which we are going to describe now. The upcoming material is heavily based on the book [17], by Bard, and the corresponding paper [18] by Bard, Courtois and Jefferson.

7.5.1 Sparsity Revisited

In Sect. 7.3.3, we showed that the sparsity of a polynomial system is a deciding factor of its solving difficulty. However, the definition of sparsity β was not properly provided. Here we begin with defining sparsity in a compact way, as we are going to need this throughout this chapter. The sparsity (β) of the system of equations, is defined as the ratio of coefficients that are non-zero, to the total number of possible coefficients. For example, in a MQ system of m equations and n unknowns over $\mathbb{GF}(2)$, sparsity can be defined as follows:

$$\beta = \frac{\kappa}{m\left(\binom{n}{2} + \binom{n}{1} + \binom{n}{0}\right)} \tag{7.2}$$

where, κ is the number of non-zero coefficients in the system.

Sparse systems, especially the overdefined once, are found to be remarkably easy to handle by SAT solvers. Typically, if $\beta \leq 1/100$, the solving is found to be way faster than exhaustive search for MQ systems. Cipher equations systems, due to practical reasons, are expected to be sparse, as sparsity implies low gate count, which is with no-doubt – desirable. Perhaps, the most fascinating advantage of SAT solvers is their relatively low memory requirements. Where the most of the standard Gröbner Basis implementations like MAGMA [37] and SINGULAR [60] crash due to excessive memory requirements, SAT solvers are found to be successful. However, the best use of SAT solvers are only possible, when certain things are being taken care of properly. In the following subsections, we are going to discuss these issues.

7.5.2 *MQ Systems on* $\mathbb{GF}(2)$

We begin our discussion here with MQ system for obvious reasons. Let us consider an ANF with n variables having m quadratic equations. Representing the system in ANF directly imply that it is in $\mathbb{GF}(2)$. One can express the ANF as the following system of equations:

$$\begin{aligned} f_1(x_1, \ldots, x_n) &= y_1 \\ f_2(x_1, \ldots, x_n) &= y_2 \\ &\ldots \\ f_m(x_1, \ldots, x_n) &= y_m \end{aligned}$$

Here each f_i is a degree-2 polynomial. Adjusting constants, it becomes sufficient to consider only $y_j = 0, \forall j$. Here, the ration $\gamma = \frac{m}{n}$ can be used to measure whether the system is overdetermined. Thus, if $\gamma = 1$, the system is exactly determined, while if $\gamma > 1$, the system is overdetermined. Let, M be the total number of monomials, where $M = \binom{n}{2} + \binom{n}{1} + 1$, as we are considering only quadratic polynomials. Let, us redefine the sparsity β, in a slightly different flavor. We define β as the expected probability of a monomial to occur or not. If $\beta << 1/2$, the system is indeed sparse. Careful inspection will show that both the current definition and the previous definitions are actually same.

The definitions in this subsection is actually a repetition of the same described in Sect. 7.3. It is however, being done intentionally to keep the readers in-sync with the contents of this section. Our next move is to describe a precise algorithm to convert such MQ ANFs to an equivalent CNF formula, solvable by state-of-the-art SAT solvers.

7.5.3 *ANF to CNF Conversion Algorithm*

The process of ANF to CNF conversion proceeds through three major steps. At the first step, some preprocessing might be performed to make the system more amenable to this conversion. Next, the system is subject to linearization with new dummy variables are added for representing monomials with degree ≥ 2. Finally, the linear system is converted to an equivalent system of clauses. At the beginning, we ignore the preprocessing step and describe the second and third steps of the process. The preprocessing step will be augmented later in this section.

Usually, constants (0 and 1) are not available in CNFs, whereas in ANF they are crucial. One can deal with this minor technicality by adding an extra CNF variable T to represent 1. The $\sim T$ then can be used to represent the 0, if required. This simple trick works fine and is being widely used, whenever such conversions are required. Once we fix the constant representation, one can directly move to step two of ANF to CNF conversion process – that is the linearization.

Polynomial System to a Linear System

Without loss of generality, let us consider only MQ systems. Every polynomial equation in an MQ ANF is a sum of linear and quadratic terms. For each monomial with degree $d > 1$ ($d = 2$ in this case), that appears in the system of equations, we add a new dummy variable. As an example, Consider, a degree-2 term, $a = wx$. The dummy variable here is a. Notice that, this is equivalent to the following three clauses:

1. $(w \wedge x) \Rightarrow a \equiv C1 : a \vee \neg w \vee \neg x = 1$
2. $\neg w \Rightarrow \neg a \equiv C2 : \neg a \vee w = 1$
3. $\neg x \Rightarrow \neg a \equiv C3 : \neg a \vee x = 1$

We can safely omit "$= 1$" terms here leading to the CNF expression $(a \vee \neg w \vee \neg x)(\neg a \vee w)(\neg a \vee x)$. This can be easily generalized for any d-degree term (monomial) to show that it results in $(d + 1)$ clauses, with total length of all the clauses, as $(3d + 1)$.

If a monomial appears more than once, the same dummy variable and the corresponding clauses can be reused. We just need to perform a frequency analysis for the monomials. Usually in large overdefined systems, where every possible monomials appear at least once, linearization with reuse of variables result in a large savings. On the other hand, for extremely sparse systems, which are not over defined as well, this is slightly pessimistic. Nevertheless, one can safely use this linearization step for all the systems at our disposal.

Linear System to CNF

After the linearization, each polynomial is now a sum of variables, or equivalently a logical XOR. Thus, consider the logical sum as $a + b + c + d = 0$. This can be represented as 8 clauses: $C1 : (a \vee \neg b \vee \neg c \vee \neg d)$, $C2 : (\neg a \vee b \vee \neg c \vee \neg d)$, $C3 : (\neg a \vee \neg b \vee \neg c \vee d)$, $C4 : (\neg a \vee \neg b \vee \neg c \vee d)$, $C5 : (\neg a \vee b \vee c \vee d)$, $C6 : (a \vee \neg b \vee c \vee d)$, $C7 : (a \vee b \vee \neg c \vee d)$, $C8 : (a \vee b \vee c \vee \neg d)$. Clearly, if we apply this trick to an l-length XOR, the total number of clauses will be $\binom{l}{1} + \binom{l}{3} + \cdots = 2^{l-1}$, which is exponential to l. So, the general process is to optimally cut the XOR into chunks or sub-sums of length c (usually 4 or 6). c is known as *cutting number*. Let as assume that it is cut into subsums of length say 4, then we get the following:

$$
\begin{aligned}
x_1 + x_2 + x_3 + y_1 &= 0 \\
y_1 + x_6 + x_7 + y_2 &= 0 \\
\vdots \quad \vdots \quad &\vdots \\
y_i + x_{4i+2} + x_{4i+3} + y_{i+1} &= 0 \\
\vdots \quad \vdots \quad &\vdots \\
y_h + x_{l-2} + x_{l-1} + x_l &= 0
\end{aligned}
$$

Assuming $l \equiv 2 \ mod \ c$ (the case for $l \not\equiv 2 \ mod \ c$ will leave the last sum requiring lesser clauses, 4 vs. 8 in the case of an even l) thus, one can have $h = \lceil \frac{l}{c} \rceil - 2$. Further, there are $h + 1$ subsums, each requiring 2^{c-1} clauses of length c each, where h is the number of dummy variables, given by $y_1, \ldots, y_h$. Finding the optimal value for the cutting number is crucial for best performance. Experimentally, it was found that $c = 6$ is the most efficient choice for the cutting number [18].

Preprocessing

The number of variables (n) in the ANF system is crucial in determining the number of dummy variables, clauses, and the number of clauses as we will show soon. Keeping the role of n in mind, one can think of a preprocessing scheme to be applied just before the linearization step. The scheme is similar to Gaussian Elimination and is due to Bard et al. [17]. The idea is to reorder the terms of any specific polynomial in and MQ system as follows:

$$x_{a_0} = x_{a_1} + x_{a_2} + \cdots + x_{a_n} + (\text{quadratic terms}) + 1$$

where the $+1$ term is optional and each $a_i \in \{1, 2, \ldots, n\}$. We add this equation to every polynomial in the system, where x_{a_0} appears, except the first one. Since SAT-solvers mostly tend to choose the most frequent variables for taking a branching-decision, x_{a_0} will not be calculated until all other variables are set. Now, if there are t such equations, then t variables will only appear in their basic definitions (not even in terms of each other) after preprocessing. So far the main system is concerned, there are now $n - t$ variables taking part in SAT branching process. This simple step is found to approximately double the SAT performance. It is worth to mention that, only the polynomials having length $\leq$4 or $\geq$10 are chosen for these replacements. This length limit is found to minimize the increase of β in the process. The reason of such behavior is attributed to the Gaussian Elimination like nature of the preprocessing algorithm. We omit further details here.

It is worth to mention that, fixing some variables a-priori, causes significant reduction to the problem size. This trick is also used frequently in the algebraic cryptanalysis community. Although, in principle, this is a violation of the basic assumptions in some sense, there could be situations in practice, where this is absolutely valid (for example, if somebody uses side channel analysis to extract some key bits before doing an algebraic attack). As a result, this trick is found to be useful in many cases, especially when the system is very large.

7.5.4 Measuring the Difficulty

Measuring the difficulty of a CNF expression is a very difficult task without solving it explicitly. However, some empirical estimations are still possible to make. Typically, the total number of variables, number of clauses and the total length of clauses may provide a rough estimation of how difficult a CNF can be. In this subsection, we will try to estimate the difficulty of the generated CNFs form its corresponding ANF expressions. In practice, this may provide a quick understanding of how judicious will it be to solve a certain polynomial system using SAT.

Without loss of generality, let us start with MQ systems. Let n ne the total number of variables and M be total number of monomials of the quadratic ANF system. Let β be the probability of occurrence of a monomial in a polynomial aka. the sparsity of the system. The cost of converting a single monomial in the linearization step is 1 dummy variable, 3 clauses of total length 7. This is required to be done for total $M - n - 1$ quadratic monomials in the system. One dummy variable will be required for the constant monomial with a clause length of 1.

The ANF to CNF conversion step after linearization will require an estimate of the expected length of each equation. For M possible coefficients, this will be equal to $M\beta$. Assuming the cutting number $c = 4$, for the time being, there will be $M\beta/2 - 1$ subsums per equation, requiring $M\beta/2 - 2$ dummy variables, $4M\beta - 8$ clauses of total length $16M\beta/2 - 32$. So in total, one should have the following:

$$\begin{aligned}
\text{Number of Variables:} &= n + 1 + (M - n - 1)(1) + m(M\beta/2 - 1)\\
\text{Number of Clauses:} &= 0 + 1 + (M - n - 1)(3) + m(4M\beta - 8)\\
\text{Length of all Clauses:} &= 0 + 1 + (m - n - 1)(7) + m(16M\beta - 32)
\end{aligned}$$

where m is the number of equations. Substituting, $m = \gamma n$, and $M = \frac{n^2}{2} + \frac{n}{2} + 1$,

$$\begin{aligned}
\text{Number of Variables:} &= \frac{n^2}{2} + \frac{\gamma n^3 \beta}{4}\\
\text{Number of Clauses:} &= \frac{3}{2}n^2 + 2\gamma n^3 \beta\\
\text{Length of all Clauses:} &= \frac{7}{2}n^2 + 8\gamma n^3 \beta
\end{aligned}$$

Furthermore, as long as $\beta > \frac{1}{m}$ the first term of each of the expression can be discarded. The other case, that is $\beta < \frac{1}{m}$ results in $(n + 1)/2\gamma$ monomials per equation, which is indeed sparse as long as γ is not quite large. One need not consider this case separately. However, the CNF difficulty measures vary with the change of the cutting numbers. Readers are referred to [17, 18] for further details on this.

So far, we have only considered MQ systems for the analysis of CNF difficulty. One should also study cubic (MC) systems for the same, as in many cases (for example the DES equations), considering the sparsity and overdefinedness, cubic equations are the most natural choice. However, cubic systems requires a lot of manual preprocessing (see the previous section and the paper by Bard [18]). The main issue with cubic systems is the number of monomials which is quite large than its quadratic counterparts. It is also a question, whether all possible monomials appear in the system or not. Intuitively, one may argue that with very large probability, many monomials will be absent. However, till the β is fixed, it is not an issue, as by definition, β is the probability of any monomial to appear in an equation – irrespective of its degree. Considering m equations, the probability of a monomial to appear in the system is $1 - (1 - \beta)^m$. Now it is interesting to compute a proper estimate for β, so that a fraction r or less of the expected monomials are present in the system, that is:

$$1 - (1 - \beta)^m \leq r$$

Since in practice, β ($< 1/2$) is a small quantity one may approximate the above inequality as $m\beta \leq r$. For practical reasons, it is only judicious to consider the cases, where actually $3/4$ or more of the monomials do not appear in the system (that is $r < 1/4$). One can now easily estimate β values form here for quadratic-cubic cases with reasonable accuracy.

For a cubic system, total $\binom{n}{3} \approx n^3/6$ cubic monomials are possible, each requiring 1 dummy variable, 4 clauses of total length 10. Note that, the quadratic monomials are still relevant here. Hence, the total number of monomials possible is $M = \binom{n}{3} + \binom{n}{2} + \binom{n}{1} + \binom{n}{0} \approx n^3/6$. The expected length of any polynomial is then $M\beta \approx \beta n^3/6$. Considering the case, where the cutting number is 4, this would require $\approx \beta n^3/12$ dummy variables and, $\approx \frac{2}{3}\beta n^3$ clauses of total length $\approx \frac{8}{3}\beta n^3$ for each of the m equations. Therefore, the cubic case is similar to the quadratic case from all aspects.

Up to this point, we have mainly discussed some important concepts and mathematical tools, mainly developed for algebraic cryptanalysis and equally relevant for AFA. With the concepts presented, it is now possible to represent any given cipher as a system of equations in ANF, and eventually as a CNF to be solved by SAT solvers. From the next section onwards, we will mainly focus on AFA. It is worth to mention that, the Sage tool [62], contains optimized python implementations for most of the algorithms discussed here. Readers are encouraged to try hands on with this tools to get a better feel of this subject.

7.6 Algebraic Fault Analysis

Algebraic cryptanalysis opened a completely new direction of classical cryptanalysis. Albeit being more systematic and formal than other state-of-the-art cryptanalytic methods, their success so far is limited. No standard cipher was ever broken with algebraic cryptanalysis, except in their reduced round versions. Even with this rather pessimistic results, algebraic cryptanalysis is found extremely useful, while combined with other attack measures – e.g. side channel analysis (SCA) or fault analysis (FA). In fact, Algebraic Side Channel Analysis is often considered as one of the most efficient side channel technique designed so far [138, 153, 184]. The main idea of ASCA is to encode a side channel trace (e.g. a power trace) in the form of equations, and to add it in the cipher equation system. It has been observed that, even for large ciphers like AES, augmentation of this leakage information solves the equation system within seconds [153], which was unsolvable otherwise. There is a significant amount of work, which address different aspects of ASCA [138, 153, 184]. We shall not discuss them here. However, the key advantage of ASCAs over other side channel techniques is that, it requires extremely low data complexity (a couple of traces are sufficient for a low noise scenario). Further, it can exploit all the data points present in a trace in a parallel manner. Although, some issues arise due to the presence of noise, one can handle that by adding some extra variables, as shown by Zhao et al. [184].

The main focus of this chapter is however, AFA, which is a relatively newer concept than the ASCA. AFA was first proposed by Courtois in [55], where it was applied on DES. It was found that, with a single 2-bit fault at the 13th round of DES, and 24 known key bits, one can recover the remaining key bits with 0.01 h, which was 10 times faster than the exhaustive search. The next notable work on AFA was due to Jovanovic et al., who demonstrated the first AFA attack on a lightweight block cipher, taking LED as an example [85]. Further progress in this direction were made by Zhao et al. [181], who demonstrated AFA on several block ciphers – mainly for the lightweight ones. It was found that AFA also works for large ciphers like AES [181] and Twofish [180]. However, the computational cost involved is extremely high (e.g. for Twofish, it took 24 h with a 280 MB attack script [180]) and the gain is not significant compared to normal DFAs. In contrast, for lightweight ciphers e.g. LED, PRESENT, Piccolo, LBlock etc., AFA was found to be capable of finding the full key within seconds [185] for most of the cases.

The main success of AFA however lies elsewhere. One can see this method as a generic framework for fault analysis attacks and security evaluation. While most of the state-of-the-art DFAs require significant amount of manual interventions, AFA only requires the cipher and additional fault equations to be constructed, which is sufficiently systematic, and can be automated easily. Taking this nature of AFA in consideration, Zhang et al. constructed a framework for the evaluation of ciphers against fault attack [180]. Our upcoming discussions on AFA will mainly follow this theme.

7.6.1 A General Fault Attack Framework

Given any cipher, there are several points where a fault can be injected. However, not all injected faults lead to a successful attack. We call a fault, capable of doing a successful attack, an *exploitable fault*. In practice, exploitable faults are not easy to figure out, given the diversity of the total fault space of a cipher. One may think of injecting a fault in various locations of the state even if the round of the fault injection remains fixed. Key schedule is another possible place, which may lead to successful attacks. Finally, one may think of modifying the round counter of a cipher, to effectively reduce security by decreasing the number of rounds. It will be shown eventually, that the equation based framework of AFA can efficiently encode all such scenarios mentioned above. Before that, we need to provide a formal model for fault attacks for the sake of mathematical analysis. Such a model should be able to precisely encode all possible fault scenarios for any given cipher. Although, there could be many ways in which such a model can be described, here we present the method due to Zhang et al. in [180].

A formal model for faults can be described as a five-tuple:

$$F = \langle X, \lambda, w, t, f \rangle.$$

Here X is an intermediate state, where the fault is to be injected, whereas, X^* denotes the faulty state after the injection of fault. Throughout this discussion, X will be called the *Position* of the fault. X_i is the unit of X which determines how the state is organized. In practice, X can be some state register of the cipher at some specific round r, or a key register or even the round counter state. Note that, the round information is closely coupled with X itself, while one considers fault injection in the state or key schedule; and one may choose to represent X as X_r^{en} or X_r^{ks} in such cases. Alternatively, if the fault is injected in the round counter, there is no need to consider the round of injection. λ here denotes the width of the state X, preferably in terms of bits.

An injected fault in our formal construction is denoted by f. While describing a fault it is mandatory to describe the fault model it follows. The fault model in this case is generalized by the fault width w. Typical values of w can be 1, 4 or 8, which denote the *bit-fault*, *nibble-fault*, and *byte-fault* models, respectively. The only remaining parameter is the t, which is referred to as the *Location* of the fault in the state X. The parameter t, depending on the fault width w, specifies the unit within X, where the fault is to be injected. Typically, for a given X, λ ($\lambda = |X|$), and w, $t \in \{0, 1, 2, \ldots, \lceil \frac{\lambda}{w} \rceil\}$. Further, depending on the scenarios, t can be known or unknown.

It is now possible to describe different fault injections in a cipher in terms of the formal construction presented above. Typically in AFA, the extra relations due to the injected faults are also encoded in the form of polynomial equations in ANF. We refer to these equations as *fault equations*. One should not be misguided by the terminology, as in the fault attack community, the same term *fault equations*, often refers to the difference equations of faults, built across the S-Boxes. In this chapter, we will use this term to refer the fault relations in ANF only. The fault equations, merged with the original cipher equation system reduces the key search complexity significantly, which eventually results in a successful attack. Following two subsections will be dedicated towards a complete ANF description of a lightweight block cipher PRESENT [33], and its corresponding fault equations as a case study. The extension of the same for other ciphers is fairly straightforward.

7.6.2 Case Study: ANF Equation System for PRESENT

We briefly recall that PRESENT is a lightweight block cipher proposed by Bogdanov et al. in CHES 2007 [33]. It has an SPN structure with 64-bit block size and 31-rounds, having a key length of 80 or 128 bits. The version with 80-bit key is widely studied and we too adapt this for our purpose. Each round of present consists of three major operations, namely: the **addRoundKey** (AK), where a 64-bit round key is XOR-ed with the state; the **sBoxlayer** (SL), which performs the substitution operation with 16 identical 4×4 S-Boxes; and the **pLayer** (PL), which is a 64-bit permutation operation. The key schedule of PRESENT-80 consists of simple bit rotation, S-Box lookup and round-counter XOR operations. The S-box and the permutation for PRESENT have already been depicted in Chap. 4 and is hence avoided here.

Equations for Round Key Addition

Round key addition results in the simplest equations of PRESENT equation system. Let, $x_0||x_1 \ldots ||x_{63}$ be a 64-bit state, $k_0||k_1 \ldots ||k_{63}$ be the round key, and $y_0||y_1 \ldots ||y_{63}$ be the output variables after round key addition. The the corresponding equation scan be represented as:

$$x_i + k_i + y_i = 0, \quad 0 \leq i \leq 63$$

S-Box Equations

The S-Box of the PRESENT can be converted to the following algebraic equations by the method described in Sect. 7.5. The input to the S-Box is denoted by $x_0||x_1||x_2||x_3$, whereas the output is $y_0||y_1||y_2||y_3$. It is worth to note that, simpler representations are indeed possible. However, From the perspective of AFA, following S-Box representation performs the best. The representation uses a degree-reverse lexicographic ordering for variables. Also, the Gröbner basis is computed to increase the number of equations. Similar representation was also utilized by Albrecht and Cid in [4]. One should notice that the representation have only a single cubic term in it. Most of the nonlinear monomials are quadratic in nature. However, it is not a complete explanation of why this S-Box representations works best for the PRESENT. In fact, recently Courtois et al. [52] proposed an even better implementation of PRESENT

S-Box which has only 14 gates in it, using intermediate variables. Such a representation may work better in this context, but not been tested yet.

$$
\begin{aligned}
y_2x_3 + y_3x_3 + x_1x_3 + x_2x_3 + x_3 &= 0\\
y_0x_3 + y_0 + y_3x_3 + y_3 + x_1x_3 + x_1 + x_2x_3 + x_2 + x_3 + 1 &= 0\\
y_3 + x_0 + x_1x_2 + x_1 + x_3 &= 0\\
y_0 + y_1 + y_3x_3 + y_3 + x_0x_2 + x_0 + x_1x_3 + x_2x_3 + x_2 + x_3 &= 0\\
y_0 + y_1 + y_3x_2 + y_3x_3 + y_3 + x_0 + x_1x_3 + x_2 + x_3 &= 0\\
y_0x_2 + y_1x_2 + y_1x_3 + y_1 + y_3x_3 + x_0 + x_1 + x_3 + 1 &= 0\\
y_1 + y_2 + y_3x_3 + x_0x_1 + x_1x_3 + x_1 + x_2x_3 + x_2 + x_3 + 1 &= 0\\
y_1 + y_2 + y_3x_1 + y_3x_3 + y_3 + x_0 + x_1 + x_2x_3 + x_2 + 1 &= 0\\
y_0 + y_2x_1 + y_2x_2 + y_3x_3 + x_0x_3 + x_0 + x_1x_3 + 1 &= 0\\
y_0 + y_1x_1 + y_1x_3 + y_1 + y_2x_2 + y_2 + y_3 + x_0x_3 + x_1x_3 + x_2 + x_3 &= 0\\
y_0x_1 + y_1x_2 + y_1x_3 + y_1 + y_2 + y_3 + x_0x_3 + x_0 + x_1 + x_2x_3 + 1 &= 0\\
y_0 + y_1x_2 + y_1x_3 + y_1 + y_2x_2 + y_2 + y_3x_0 + y_3x_3 + x_0x_3 + x_1 + x_2x_3 + x_3 &= 0\\
y_0 + y_1x_2 + y_1 + y_2x_0 + y_2 + x_0x_3 + x_1 + x_2 + x_3 &= 0\\
y_0 + y_1x_0 + y_1x_3 + y_2 + y_3 + x_0x_3 + x_0 + x_1x_3 + x_1 + x_2x_3 + x_3 + 1 &= 0\\
y_0x_0 + y_0 + y_1x_3 + y_1 + y_2x_2 + y_3 + x_0 + x_1x_3 + x_3 &= 0\\
y_0 + y_1x_3 + y_1 + y_2y_3 + y_2 + y_3x_3 + y_3 + x_0x_3 + x_0 + x_2x_3 &= 0\\
y_1y_3 + y_1x_2 + y_1x_3 + y_1 + y_2x_2 + y_3x_3 + y_3 + x_0x_3 + x_1x_3 + 1 &= 0\\
y_0y_3 + y_0 + y_1x_3 + y_2 + y_3x_3 + x_0x_3 + x_1x_3 + x_1 + x_3 + 1 &= 0\\
y_0 + y_1y_2 + y_1x_3 + x_0x_3 + x_0 + x_1x_3 + 1 &= 0\\
y_0y_2 + y_1 + y_3 + x_3 + 1 &= 0\\
y_0y_1 + y_0 + y_1x_3 + y_1 + y_2 + x_0x_3 + x_2x_3 + x_2 + x_3 + 1 &= 0\\
y_1x_2x_3 + y_1x_2 + y_3x_3 + y_3 + x_0x_3 + x_0 + x_1x_3 + x_1 + x_2x_3 + x_2 &= 0
\end{aligned}
$$

Equations for the Permutation Layer

The easiest way to represent the permutation layer of PRESENT is to use the following equation, where x_i, and y_i $(0 \leq i \leq 63)$ denote the input and output variables respectively.

$$x_i + y_{P(i)} = 0, \quad 0 \leq i \leq 63$$

The Complete Equation System

The complete equation representation of PRESENT requires new set of variables for each round. Also, one need to write down the key schedule equations, otherwise the key variables across rounds will be considered to be independent and will prohibitively increase the complexity of search. We do not provide the key schedule equations here and leave them for the readers as an exercise. As, for the datapath equations, each round will be connected to the next round variables by means of the permutation layer equations. The complete equation system consists of 13642 equations over 4172 variables. It is worth to mention that, while constructing the equation system for a single plaintext-ciphertext pair, one must add them along with the equations. If there are multiple pairs of plaintext and ciphertext, one must assign new set of variables and equations for each of them. Recently, authors in [181] claimed that using the decryption equations instead of the encryption equations, makes AFA run faster. This claim has been supported experimentally in some cases. However, for many cases (like the PRESENT) encryption equations work equally well. So, it still remains an open question that which is the optimal choice.

Once the cipher equation system is constructed, the next obvious step is to represent the faults as ANF equations. This part is exclusive for AFA, and dramatically reduces the search apace complexity for the key. From the next section onwards, we analyze the effect of faults on the cipher.

7.6.3 Construction of Fault Equations

Construction of fault equations is quite generic in AFA than that of DFA. Given a fault $F(X, \lambda, w, t, f)$, we start by adding extra variables to represent the faulty values in the system. Typically, for a fault injection at round r of a r_{max} round cipher, one need to assign additional variables for all rounds following r (also including r itself) to represent the faulty values. Let Y denote the faulty state at round r, whereas X denote the correct state. If the size of the state is λ-bit, one can represent them as $Y = y_1||y_2||\ldots||y_\lambda$ and, $X = x_1||x_2||\ldots||x_\lambda$, respectively. Now given the fault width w, one could have $m = \frac{\lambda}{w}$ possible locations for the fault injection. Note that we are talking about the injection of a single fault instance at a time in the state.

Let $Z = z_1||z_2||\ldots||z_\lambda$ denote the differences between X and Y. Indeed, they can be represented by means of the following equations:

$$z_i + x_i + y_i = 0, \quad 1 \leq i \leq \lambda \tag{7.3}$$

Given the m possible locations for the fault injection, we divide Z in m chunks $Z_1||Z_2\ldots||Z_m$ as follows:

$$Z_i = z_{w\times(i-1)+1}||z_{w\times(i-1)+2}||\ldots||z_{w\times i}, \quad 1 \leq i \leq m \tag{7.4}$$

In a practical scenario, the exact location of t may or may not be known to the adversary. Depending on this knowledge, one may have a different representations of the fault equations. If the t is known, then Z can be denoted as follows:

$$Z_i = 0, \quad 1 \leq i \leq m, \quad i \neq t \tag{7.5}$$

and

$$u_t = (1 + z_{w\times(t-1)+1})(1 + z_{w\times(t-1)+2})\ldots(1 + z_{w\times t}) = 0, \tag{7.6}$$

where u_t is a bit variable to represent that Z_t is faulty. Using Eqs. (7.5) and (7.6), fault is represented with $w + 1$ variables and $w(m + 1) + 2$ CNF equations.

The second case arises, while the exact location of t in the target round r is unknown to the adversary. In this case, we introduce variables u_i $1 \le i \le m$, to represent the fault status of each Z_i chunks.

$$u_i = (1 + z_{w\times(i-1)+1})(1 + z_{w\times(i-1)+2}) \ldots (1 + z_{w\times i}), \ 1 \le i \le m \tag{7.7}$$

If $u_i = 0$, Z_i will be the variable associated with the w-bit fault. With the assumption that, one and only one Z_i chunk can be faulty, there should be only a single u_i which is 0. This constraint can be represented by the following equations:

$$\begin{aligned} &(1 + u_1) \vee (1 + u_2) \vee \cdots \vee (1 + u_m) = 1 \\ &u_i \vee u_j = 1, \ \ 1 \le i < j \le m \end{aligned} \tag{7.8}$$

Equations (7.7) and (7.8) results in $m(w + 2)$ variables and $m(2w + 0.5m + 1)$ CNF equations. The equations can be parameterized by the faults of w, m and λ. Note that, the value of the fault f is not required at any point of equation construction. For PRESENT, typical values for λ, w, and m can be taken as 64, 4 (nibble fault model) and 16 respectively. Bit and byte fault models are equally relevant in this situation. It is quite easy to observe that, the complete equation construction process is totally generic, which makes AFA a powerful tool for automatic fault analysis.

7.6.4 The AFA Procedure

A successful AFA often requires more than one faulty ciphertexts to succeed. The main idea is to reduce the key search space size to a very small value. Typically, with each injection, the search complexity reduces to some extent. Although, the reduction can be analyzed mathematically, it is not required for a successful master key recovery with AFA. Later in this section, we shall show an automated computational way of estimating the search space size.

Algorithm 8 presents the complete pseudo-code of AFA. It is adapted from [180], with some minor modifications. For the time being, we consider that the fault is injected on the datapath of the encryption algorithm. Similar procedures can be constructed for the faults in the key schedule and round counters. Referring to Algorithm 8, one begins with N number of w bit faults to be inserted in the rth encryption round. The location t may be unknown depending on the scenario. The output of the algorithm is the time to find out the master key(t_{sol}). Note that, depending on the scenario, the algorithm may or may not be able to solve the problem instance within reasonable time. The standard solution for this scenario is to set a reasonable timeout t_{out}. In case the algorithm fails to figure out the key it outputs the $t_{sol} = t_{out}$. Note that, the Algorithm 8 is constructed as a verification mechanism and the key is generated internally. However, one may easily supply the plaintexts and ciphertexts, encrypted with some fault instance and unknown key, to this algorithm to construct a true attack algorithm.

Algorithm 8 The AFA Procedure for the Datapath Faults

Input: N, r, w, b_t, t_{out}
Output: t_{sol}
`RandomPT` ($\mathbb{P}$)
$\mathbb{K}$ = `KS` (K, L)
for $rc = 1$ to r_{max} **do**
    `GenKSRdES` (rc, K_{rc+1}) ▷ #1
end for
for $i = 0$ to $N - 1$ **do**
    C_i = `Enc` (P_i, K)
    for $rc = r - 1$ to $r_{max} - 1$ **do**
        `GenEnRdES` $(X_{rc+1}, X_{rc}, K_{rc+1})$ ▷ #2
    end for
    `GenInputES` (C_i)
    C_i^* = `InjectFault` (`Enc` $(P_i, K), X_r)$
    for $rc = r - 1$ to $r_{max} - 1$ **do**
        `GenEnRdES` $(X_{rc+1}^*, X_{rc}^*, K_{rc+1})$ ▷ #3
    end for
    `GenInputES` (C_i^*)
    `GenFaultyES` $(f = X_r + X_r^*)$ ▷ #4
end for
`RandomPT` (P_v)
C_v = `Enc` (P_v, K)
for $rc = 0$ to $r_{max} - 1$ **do**
    `GenEnRdES` $(X'_{rc+1}, X'_{rc}, K_{rc+1})$ ▷ #5
end for
`GenInputES` (P_v, C_v) ▷ #6
t_{sol} = `RunAFA` () ▷ #7

In Algorithm 8, $\mathbb{P}$ and $\mathbb{K}$ denote the plaintext and ciphertext spaces, respectively. `KS` and `Enc` denote the key scheduling and encryption functions, respectively. `RandomPT` generates one or more random plaintexts and `InjectFault` induces one fault. The functions generating the equation sets are prefixed with `Gen` and suffixed with `ES`. During the attack, the adversary chooses N pairs of plaintexts/ciphertexts and starts constructing equations. First, the equations for the key schedule are built (#1 in the algorithm). Next, each of the N encryptions are subjected to a fault injection and corresponding correct and faulty equations are constructed, starting from the round of injection (r), along with the corresponding fault equations at the round of injection (see #2, #3, and #4 in the algorithm). This system, if solve is supposed to return multiple solutions, all except one of which are wrong. The key candidates are from the reduced key search space after the injection of faults. Depending on the parameters of the injected faults, the number of key candidates will vary. In as we shall see later in this section, number of key candidates may vary from 1 to 2^s, where $s \leq \lambda_s$ can be any number up to the key size λ_s. In the works case, one may expect no significant reduction in the key search space.

To eliminate the spurious solutions automatically, a validation pair of plaintext-ciphertext (P_v, C_v) is added to the equation system. One need to construct the a complete encryption/decryption equation set, with the validation pair. The target of this step is to pinpoint the correct key and return it. However, one should note that, if the size of the remaining key space after fault injection is significantly large, the solution time with this validation set included may increase prohibitively. Once the equation system is constructed, it is converted to CNF and is fed to a SAT solver. The most popular choice in the AFA community is the CryptoMiniSAT [168], which is specially tailored to handle the XOR clauses. As we see, XOR clauses are quite common in cryptographic problems, which makes CryptoMiniSAT a perfect candidate for the purpose.

Given the general procedure for an AFA attack, let us investigate the effect of it on the datapath of PRESENT. A comprehensive analysis of the result we present, can be found in [180]. Here we just provide a brief summary of that. Note that, only PRESENT-80 was considered for attack. Also, attacks at the 28th and 29th round were studied for different fault widths. It has been found that, the injection in 28th round of PRESENT is more efficient than that of the 29th round. In fact, 2 fault ($w = 1$, bit fault) injections can recover the key within three minutes for most of the cases in the 28th round. It was further assumed that t is known to the adversary. For other fault models $w = 4, 8, 16, 32$, the performance degrades. Performance was the worst for the 32-bit fault. For nibble faults, the attack at the same round may take up to 9000 s for some cases. In fact, for some cases it is not possible to solve the system even with a large timeout. Nevertheless, AFA is able to detect several attacks on the 28th round of the cipher with less number of fault injections, which was previously beyond the scope of manual effort as shown in [186]. In fact this trend is quite general for most of the lightweight block ciphers, which gives AFA an extra foothold as an automated tool for finding newer attacks.

The attacks in the 29th round of PRESENT are indeed possible in principle. However, the AFA procedure is not able to extract the key in these cases due to incomplete diffusion of the faults. One possible way is to increase the number of fault injections. However, it will require a large number of faults to be inserted. Another alternative is to compute the complexity of the key space after each fault injection. AFA adapts the second procedure, and it will be discussed in Sect. 7.6.6. In the next subsection, we shall see the effectiveness of the AFA procedure for different fault locations.

7.6.5 Beyond the Data-Path: Attacking Key Schedule and Round Counters

Fault attacks are not limited to the datapaths of the ciphers. The key schedules often become target for an adversary. Another easy way to attack the cipher is to reduce of increase the number of rounds by corrupting the round counters. In this section, we shall provide an conceptual overview of these attacks in the context of AFA. The details of these attacks on the lightweight block cipher *LBlock*, can be found in [180].

An attack on the key schedule is quite similar to the attack on the datapath. Here the adversary needs to inject a fault in the key schedule at some round r. The only difference is that, now he has to construct the equations for the faulty computation using the new variables for the faulty key. Referring to Algorithm 8, we need to construct the faulty key schedule equations just after the construction of correct round equations in lines 8–11. Next, the faulty round equations will be constructed – this time using the faulty key bits. Rest of the things will remain same. The experiments on *LBlock* in [180], reveals that, usually the number of injections and the time complexity is more than that of a datapath attack. Further, it was shown that the attack can work at deeper round of the key schedule, which was previously beyond the scope of the manual attempts.

The second possible attack scenario is to modify the rounds of the block cipher. Practically, this can be done in two ways – by modifying the round-counter rc or by modifying r_{max}. We begin with the second case here. The scenario can be modeled just by modifying the value of r_{max} to $r^{'}_{max}$, during the generation of the faulty round equations (line 12–15 in Algorithm 8). Now it can be observed that if $r^{'}_{max} \geq r_{max}$, no useful information can be obtained from the attack. However, if $r^{'}_{max} < r_{max}$, the attacks can succeed. Typically, for small values of $r^{'}_{max}$, (3 or 4), the attack on LBlock, returns the key within seconds. This is quite obvious as we are attacking a reduced round version of the cipher in this case. Similar results can be obtained for PRESENT as well. However, for large values of $r^{'}_{max}$, the attack may not succeed.

Next, we consider modifying the rc value of the cipher during its execution. Several cases may arise here. We describe the three main cases from [180]. In the first case, some intermediate round is repeated, but total number of rounds remain the same as r_{max}. For LBlock the attack was successful within reasonable time. The second case, skips some of the intermediate rounds. This is a quite practical scenario as one can observe for *instruction skip attacks*. As per [180], the attack also succeeds in this case for LBlock cipher. The third case is to continue the computation beyond r_{max} round, which is same as modifying r_{max} to a larger value. As pointed out previously, this is not an useful case, and does not help the adversary by revealing any information.

From the discussion above, it is evident that AFA can model and solve most of the practical fault scenarios discovered till date. However, one should note that, it may not be possible to extract the key in all possible cases for any given cipher. The recovery typically depends upon the complexity of the cipher and the nature of the diffusion layers. A standard trend in the fault attack community is to calculate the attack complexity, even for the cases where the complete key extraction is not feasible. The next section will provide the way by which the algebraic framework can be utilized for this purpose. This will make the algebraic more complete from a tool perspective.

7.6.6 Analyzing the Attack Complexity

In general, the complexity of the fault attack is determined by the size of the remaining key space after the injection of faults. It is possible to find the bounds on the remaining key space through mathematical analysis, as it is often done in traditional DFAs. However, for AFA, we have an automatic procedure. The idea is to count the number of solutions of the equation system after the injection of faults in the cipher. Although, its seems to be pretty much straightforward, the problem is #P-complete in nature, which is even difficult than the NP-Complete class of problems. Figuring out some approximation seems to be the only way to deal with this scenario. In [180], Zhang et al. proposed an approximate information metric for the complexity calculation, which is a crude but acceptable approach for this purpose. In this subsection, we present this approach in detail along with the results for PRESENT. Interested readers may refer to the paper by Zhang et al., for detailed experimental study.

Approximate Information Metric

The approximation information metric ($\phi(K)$) basically calculates the conditional entropy of the secret key after N fault injections. The remaining key search space is basically $2^{\phi(K)}$, which, if reasonable returns the secret key in practical amount of time. However, in a typical fault injection scenario, the number of solutions may be too large to enumerate. As a remedy, κ key bits can be supplied a-priori, to reduce the count. The metric is represented in a logarithmic scale. Before describing the metric mathematically, let us first describe the equation system we require for this purpose. The setup is almost similar to that described in Algorithm 8. Only the validation equations are removed (line 19–24). This enables the system to have multiple solutions. The CryptoMiniSAT solver, which can be set to run in multiple solution mode, is then used to count the number of solutions. Following the convention of [180], we call this new equation system and corresponding solving mechanism as *Mode-B*, whereas the methodology presented in Algorithm 8 is termed as *Mode-A*.

Algorithm 9, adapted from [180], presents the pseudo-code for the calculation of $\phi(K)$. Referring to Algorithm 9, len denote the key length and κ denote the number of guessed secret bits fed into the system. To estimate $\phi(K)$, κ is reduced from a larger value to the smaller value, until 0. Let $\eta(\kappa)$ denotes the number of solutions for a given κ. If the number of solutions is too large, (typically larger than 2^{18}), it becomes computationally expensive to enumerate all the solutions. The remedy is to put a threshold $\tau = 2^{18}$. In Algorithm 9, `GenerateAFAES` generates the equation system with N faults injected as described for *Mode- B*. `GenKnownKeySet` sets the known key values in S_k, that is S_k is initialized with all the known secret key values. `FeedRandKeyBits` feeds κ secret key bits in the equation system, whereas `RemoveRandKeyBits` remove a random key bit. Next, the `RunAFAModeB` is called which runs the CryptoMiniSAT solver in multiple solution mode. The solutions are counted next and the information metric is calculated using the formule described in line 9 or 13, depending on the number of solutions.

Although, the calculation of the information metric seems quite tedious, it provides a reasonable solution to the complexity calculation problem. Here we just mention the

Algorithm 9 Estimation of $\phi(K)$

Input: len, N, τ
Output: $\phi(K)$

GenerateAFAES (N)
GenKnownKeySet (S_k)
for $\kappa = len; \kappa > -1; \kappa --$ **do**
  FeedRandKeyBits (S_k)
  RemoveRandKeyBit (S_k)
  RunAFAModeB ()
  CalcSolutionCount ($\eta(\kappa)$)
  if $\eta(\kappa) \geq \tau$ and $\kappa > 0$ **then**
    $\phi(K) = \kappa + log_2(\eta(\kappa))$
    break
  end if
  if $\eta(\kappa) < \tau$ and $\kappa == 0$ **then**
    $\phi(K) = log_2(\eta(\kappa))$
  end if
end for

results of the metric for the fault attacks on PRESENT. Typically, for the attacks on 28th round, a single fault injection reduces the $\phi(K)$ value to 30 for 35% of the cases. This implies that two fault injections will be sufficient and that happens in reality as well. On the other hand, a single fault injection in the 29th round, reduces $\phi(K)$ to 64 for almost 40% of the cases. Such a situation may not enable one to recover the master key though; but it provides a good estimate regarding the exploitability of the fault.

7.6.7 AFA on Different Ciphers

Several successful demonstrations of AFA can be found in literature [180, 181, 185]. Most of the cases attacks were done on lightweight block ciphers. The primary reason for this is the prohibitive complexity of AFA for them. For example, the optimal DFA on AES usually takes 50 min, whereas for AFA, it takes around 10 h. However, for the lightweight ciphers and even for some so-called large ciphers like DES, AFA works faster than DFAs. The main strength of AFA is however its capability of exploiting complex relations in the form of cipher equations, which enables the discovery of newer attacks with lesser number of faults.

In [181], Zhang et al. reported AFA attacks on Piccolo, AES, DES MIBS-64 and LED-64. For Piccolo, faults were injected at two different rounds. It was reported that with total 3 fault injections (nibble/byte fault), the master key can be recovered within maximum 1800 s. The optimal attacks on AES took 10 h to complete. For MIBS-64 the attack took 1100 s with a nibble fault at 29th round. For DES, the best result was reported for a single bit fault at 11th round which took around 3000 s to complete. However, it is worth to mention that, these are only some possible attack instances and given the diversity of the fault space, better attacks can be found. For LED-64, the standard AFA attempt took around 15 h to complete with a single nibble fault at 30th round. However, with the addition of some manually found relations [185], the complexity reduces to 180 s for most of the cases.

In [180], a comprehensive analysis on LBlock cipher with various attack scenarios were presented, which shows that LBlock can be attacked even with single bit faults within minutes. Calculation of approximate information metric were also provided for the LBlock at different which provides some more potential exploitable fault cases, which may take more time to discover the key. We would not go into the details of it here. Further study on DES was performed and it was found that the complexity of the AFA attack depends on the location of the fault injection, even within same round. Approximate information metric values were also provided supporting this claim. The most complex attack instance of AFA was provided in this chapter in terms of attacks on the Twofish [163] block cipher. Twofish was subjected to only a single DFA attempt in [7]. Although, the AFA was not able to find any newer attack, it showed significant in terms of the number of faults.

7.7 Discussions

Genericness of AFA attacks provides powerful means of automating fault based cryptanalysis. However, there are several open issues to be answered. Alternatives to AFA has also been proposed recently in several contexts for finding vulnerability in cryptographic implementations. In [21], Barthe et al. presented a mechanism to find exploitable faults in public key ciphers like RSA and ECDSA using a variation of program synthesis techniques. A more relevant work can be found in [95], which tracks the fault propagation paths inside a cipher using a coloring based abstraction. Each color represents a new variable. The complexity calculation here can be done through a simple static analysis. Results were provided for large block ciphers like AES, SMS4 and CLEFIA. The main issue with this work is that it is somewhat over simplistic and cannot exploit complex relations arising due fault propagation, which is the main source of power for AFA. Further, it cannot deal with typical ciphers like Twofish. Another notable work is [38], which provides the most optimal attack tool for the AES. Key extraction of AES with this tool, can be done within a second. However, being too specific to a cipher, this tool is not that useful in the present context, where we are mainly looking for generic tools.

One major complain against the AFA is that it does not perform well on so-called large cipher structures. Although, in [180], some results were provided for Twofish, which is considered to be an extremely complex cipher, it took around 24 h for the AFA tool to get the results. Given the diversity of the fault space and injection mechanisms for any given cipher, this time is surely a bottleneck for the discovery of the complete fault space. Future research on AFA should try to get rid of these performance issues for large block ciphers.

7.8 Summary

In this chapter, we have provided a comprehensive survey of the Algebraic Fault Attacks, which have recently gained significant attention form the fault attack research community due to its simplicity, and generic nature. We tried to keep the discussion as comprehensive as possible and skipped some mathematical details (e.g. discussions on Gröbner basis) to maintain that. Also, we did not provide comprehensive results of the attacks which are already there in the papers. The target was to provide mainly the technical details, which should enable the readers to do the attack by themselves after reading this chapter. Finally, we have provided almost exhaustive literature review of the body of work on AFA for block ciphers.

Automation of fault based cryptanalysis is a need of the present day, and an active area of research. While designing countermeasures, one must provably secure the cipher against every member of the exploitable fault space. Although, traditional fault tolerant design methodologies were proposed in this context by several researchers, they are indeed not optimal without the knowledge of the exploitable fault space. Further, for lightweight ciphers, which are meant for resource constraint environments, traditional countermeasures could be an overkill, as it often takes each fault to be equally probable. The knowledge of the fault space is thus an invaluable resource for system designers. From a practical perspective, automation will be inevitable as it is not reasonable to provide man hours behind analyzing a cipher against each and every attack. Rather the task should be automated and should be embedded within the synthesis tools, so that they can generate the proper countermeasures for each target implementation.

Another area, that has not received significant attention until now is the automation of fault intensity analysis (FIA) [69]. Unlike fault analysis attacks, which are mainly algorithm specific, FIAs are dependent on the physical properties of the device. Although, recent works have focused on designing formally provable countermeasures for such attacks [46, 126, 152], no framework is available currently for the vulnerability analysis, as per the knowledge of the author. This is another potential area of future work.

Part IV
Countermeasures Against Fault Analysis Techniques

Chapter 8
Classical Countermeasures Against Differential Fault Analysis

Sikhar Patranabis and Debdeep Mukhopadhyay

Abstract In the preceding discussion, we have presented to the readers a variety of fault attacks on a wide range of cryptographic algorithms across hardware and software-based implementations. It is therefore established, by now, that fault attacks are a serious threat to the security of hardware implementations, and consequently, sound countermeasures must be designed to tackle such threats. In this section, we present a classical approach popularly employed to resist such attacks, namely detection. Since most of the fault attacks described so far exploit transient faults during algorithm execution rather than permanent faults in the target device, the countermeasure strategy is to detect any such occurrence of faults during every execution of the cryptographic algorithm.

8.1 Introduction

Detection mechanisms against fault attacks, also referred to as concurrent error detection (CED) techniques, rely on the identification of fault occurrences via specialized circuits using some form of spatial, temporal or even information redundancy. For example, a simple detection methodology could be to duplicate the computation steps of the cryptographic algorithm at various points during ts execution, and comparing the results of the two computations to detect a fault. If a fault is detected, the ciphertext is either suppressed, or is randomized so as to render it unexploitable for an adversary. We now explore some popular detection-based countermeasure techniques against DFA.

S. Patranabis (✉) · D. Mukhopadhyay
Secured Embedded Architecture Lab, IIT Kharagpur, Kharagpur, India
e-mail: sikhar.patranabis@iitkgp.ac.in

S. Patranabis and D. Mukhopadhyay (eds.), *Fault Tolerant Architectures for Cryptography and Hardware Security*, Computer Architecture and Design Methodologies, https://doi.org/10.1007/978-981-10-1387-4_8

8.2 Spatial Redundancy

Spatial redundancy is perhaps the simplest and most intuitive detection-based countermeasure against DFA. It simply duplicates the algorithmic computation, with checks performed at regular intervals between the original and redundant computations to identify any faults, if present. An example of a spatial redundancy-based countermeasure is presented in [86], where an implementation of AES duplicates the original hardware circuit to detect faults that may have been injected during execution. As shown in Fig. 8.1, the countermeasure idea is to inter-mix state bytes between the original and redundant implementations in two separate pieces of hardware. Duplication of hardware results in an area overhead of approximately 200%, while the performance overhead is minimal as the original and redundant implementations operate in parallel within the same number of clock cycles.

An alternative technique [136] that reduces the hardware overhead via partial-redundancy is demonstrated in Fig. 8.2. The idea here is to duplicate one of out of four S-Box operations, and to perform a total of two checks per S-Box across the 10 rounds of an encryption cycle. Although this technique reduces the hardware overhead to 26%, it also suffers from a reduced fault coverage potential of 25%, since only one in four S-Boxes is effectively protected by this technique in every clock cycle.

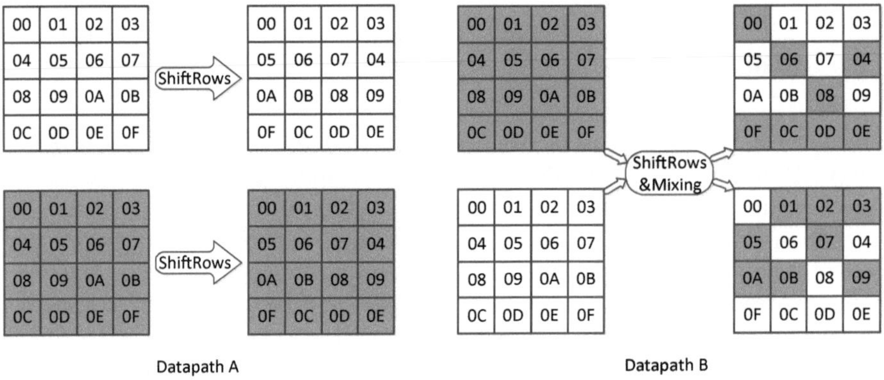

Fig. 8.1 Hardware datapath mixing: spatial redundancy

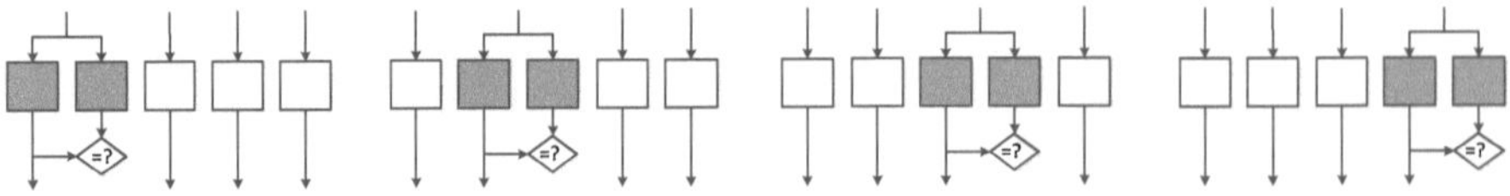

Fig. 8.2 Partial spatial redundancy

8.3 Temporal Redundancy

As opposed to spatial redundancy that duplicates each operation using additional hardware operating in parallel, temporal redundancy performs each operation twice iteratively using the same hardware, and then compares their results. Naturally, this mechanism has much lesser hardware overhead as compared to spatial redundancy techniques, but potentially suffer from a throughput reduction by around 50%. In [116], the authors tackle this mechanism using a double-data-rate (DDR) mechanism for protecting AES. The proposed implementation comprises of a pipelined AES datapath logic partitioned into two classes - with non-adjacent stages in the pipeline being driven by two opposite clock signals. This design allows halving the number of clock cycles required per round, albeit at the cost of some impact on the overall maximum clock frequency, hence the name *double-data-rate*. The additionally available free cycles are now used to recompute each round additionally on the same operand, followed by a comparison of the outputs of the original and redundant round computations for fault detection. The design achieves maximal fault coverage at a hardware cost of around 36%. A major disadvantage of this technique is the feasibility of retaining efficient implementations as technology scales.

An alternative strategy for better throughput is the *sliding-based* pipelined block cipher design proposed in [151] (see Fig. 8.3). The idea is to save clock cycles by using different pipeline stages perform the original and redundant computations, respectively, giving the approach a *sliding* characteristic. Suppose that the pipeline has a total of r stages, and that the first stage is currently performing a given computation at some point of time. In a normal unprotected encryption mechanism, the rth stage of the pipeline would be idle at this stage, waiting for the output from the $(r-1)$th stage. However, in the *sliding-based* countermeasure, the rth stage of the pipeline performs a redundant version of the same computation in parallel, and the outputs of the two pipeline stages are then compared to check for the presence of a fault. Similarly, when the original computation shifts to the second stage of the pipeline, the first stage of the pipeline (which was supposed to be idle in the normal circumstance), now performs the corresponding redundant operation in parallel. This continues till it is the turn of the rth stage of the pipeline to perform the original computation, while the corresponding redundant computation is performed by the $(r-1)$th stage in parallel, and the final comparison is performed between the outputs of these two stages. In some sense, this is similar to spatial redundancy, since the original and redundant computations occur in parallel. However, since the computations occur in two different stages of the pipeline, each pipeline stage can use its own dedicated hardware component, and the only additional hardware is for the comparators and the multiplexers. In case of AES-128, this design strategy is found to retain a throughput of up to 90% as compared to the original unprotected design, while incurring a hardware overhead of only 2.3%.

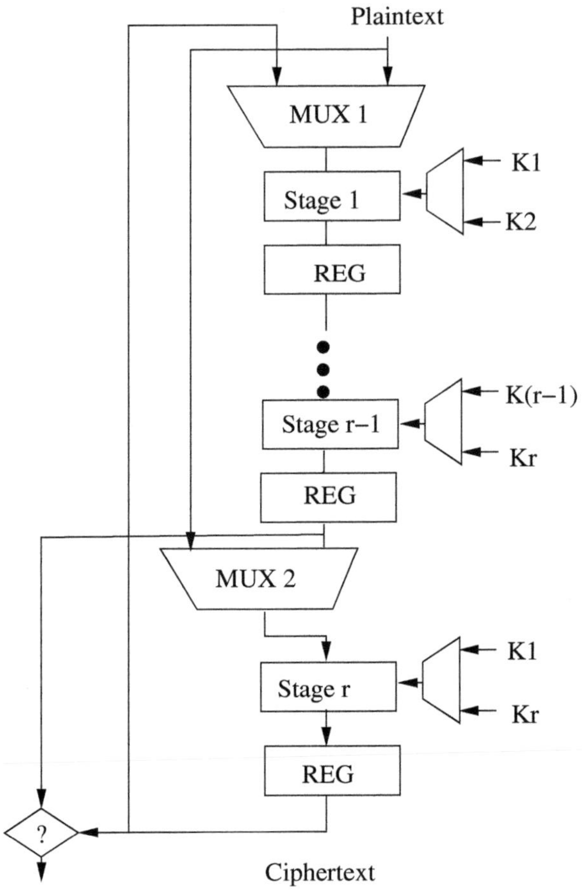

Fig. 8.3 Slid CED architecture for time redundancy

8.4 Information Redundancy: Code-Based Detection Techniques

Given the high overhead of the spatial and temporal redundancy-based techniques, an alternative approach proposed by the designers is to propose information redundancy based techniques that usually have lower overhead. Such techniques usually rely on the use of error-detection codes. Usually, in these techniques, a set of check-bits are generated that propagate through the encryption algorithm along with the plaintext message, and can finally be used to validate the fault-free nature of the ciphertext so generated. A large number of such techniques have been studied in the literature [27, 93, 94, 127–130], and we present to the readers a few of them here.

8.4.1 Parity-Based Redundancy

As the name suggests, parity-based redundancy uses a set of one or more parity bit To detect the occurrence of faults during the encryption algorithm. The first such technique to be proposed in the context of AES-128 [178] used a single parity bit for the entire state matrix. The input parity bit for each round is transformed into a corresponding output parity bit, which is used to validate the correctness of the intermediate state after every cipher round. Since the ShiftRow and MixColumn operations are parity-preserving by design, there is no explicit transformation of input parity to output parity specifically for these operations. The parity transformation owing to the AddRoundKey operation can be easily pre-computed as a function of the round key. It remains to transform the parity bit across the S-Box layer in each round. To achieve this, each of the 16 S-Boxes are transformed into 8×9 mappings from their original 8×8 mapping-based forms. The additional bit in the output of each S-Box is used to modify the input parity bit into the output parity bit, for each state byte. Although this technique has a hardware overhead of only 22.3%, the corresponding fault coverage is also low - around 48–53% on an average against randomly injected faults. Although this technique has a hardware overhead of only 22.3%, the corresponding fault coverage is also low - around 48–53% on an average against randomly injected faults.

The above technique, often referred to as **Parity-1** due to its use of a single parity bit for the entire state matrix of AES, was extended to **Parity-16** in [27]. This technique uses 16 parity bits (one per byte of the state matrix) as opposed to a single bit for the entire state. A realization of this technique for 128-bit iterated hardware implementation of AES using LUT-based S-Boxes is demonstrated in Fig. 8.4. There are two main changes made in this technique as opposed to Parity-1:

- The S-Boxes are now 9×9 as opposed to 8×9, such that all S-Boxes that receive inputs associated with invalid parity bits are automatically directed to output values

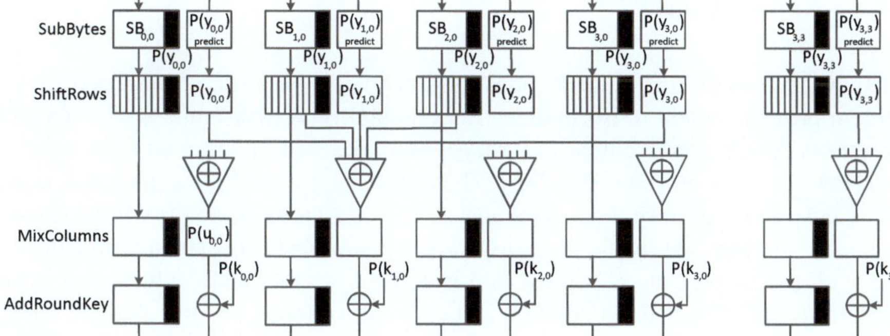

Fig. 8.4 Parity-16 CED

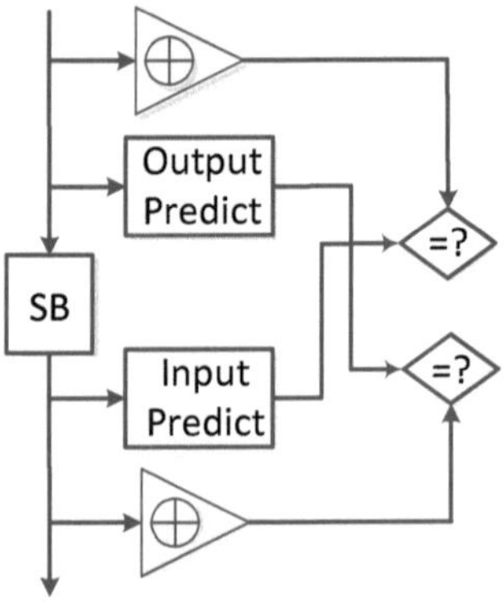

Fig. 8.5 Parity-32 CED

associated with incorrect output parity bits. This allows detecting errors in the read-only memory (ROM).
- The MixColumn operation now affects the parity of individual bytes; consequently, the technique uses additional parity-prediction units for the MixColumn operation.

The other operations, namely ShiftRow and AddRoundKey, are handled as before. This technique incurs a hardware overhead of 88.9% because of the LUT size for the S-Boxes is doubled, while the throughput is 67.86% of the original AES-128 implementation.

The final member of this family of countermeasure is **Parity-32** [135], as shown in Fig. 8.5. The technique uses one parity bit per byte of the AES state matrix for all operations other than SubByte, for which it uses *two* parity bits instead of one for improved fault tolerance. The additional input-output parity bit pair is predicted, and compared with their respective computed counterparts. The implementation reported in [135] is highly optimized, with a hardware overhead of only 37.35%, and a high fault coverage of 99.2%.

8.4.2 Robust Codes

A disadvantage of the use of linear codes, as in any parity-based detection mechanism, is that it provides a non-uniform detection probability against faults with even and odd parity. As an improvement to parity-based detection, alternative robust code-based designs were introduced in [89]. These designs use non-linear error detection codes, and provide uniform fault coverage, as long as all possible faults within a given fault space are injected by an adversary with equal probability. The architecture of AES with robust protection is presented in Fig. 8.6. In this architecture, two extra units are needed. One is the prediction unit at the round input, and it includes a linear predictor, a linear compressor, and a cubic function. The other one is the comparison unit at the output of the round, and it includes a compressor, a linear compressor, and

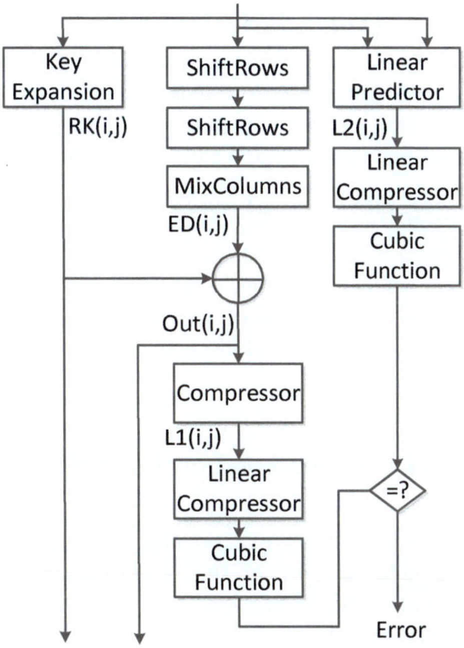

Fig. 8.6 Robust code-based CED

a cubic function. This architecture protects the encryption and decryption as well as key-scheduler module.

The linear portion of the design comprises of a linear predictor and a linear compressor. These are designed to produce a 32-bit output, that has a linearly related to the output of a given AES round, as depicted in Fig. 8.6. Note that in comparison to the design for an original AES round, the linear portion is highly compact. The round function is significantly simplified by XOR-ing the bytes in the same column, while the linear portion completely ignores the effect of the MixColumn. This in turn does with the need for costly multiplication primitives in both the MixColumn and InvMixColumn modules. The non-linear cubic portion cubes the input in $GF(2^r)$ to produce the r-bit output. In the comparison unit, the compressor and the linear compressor are designed to generate a 32-bit output from the 128-bit round output. The generation follows the same two-stage process of first XOR-ing the bytes in the same column of the output, followed by the application of the cubic function to generate the r-bit output. This output is then compared with the output from the prediction unit. Robust codes provide a fault coverage of $(1 - 2^{-56}) \times 100\%$, and impose a hardware overhead of 77%.

8.5 Hybrid Redundancy

As the name suggests, hybrid redundancy is a combination of one or more of the classical redundancy strategies described above. The foremost example of such a combination is to make each operation redundant by comparison of its input with the output its corresponding inverse operation. The redundancy could be at various granularities, ranging from the algorithm level [92], to the round level [91], and even at the basic operation level [90]. Under the assumption that complete encryption device operating in the ECB mode consists of both encryption and decryption modules, this approach yields low-cost and low-latency systematic concurrent error-detection (CED) techniques.

8.5.1 Algorithm-Level Hybrid Redundancy

Consider, for example, the algorithm-level CED for AES described in Fig. 8.7, where the plaintext, which is the input to the encryption algorithm, is compared against the output of the decryption algorithm. Since AES uses the same set of round keys for both encryption and decryption, they can be generated a priori, stored in the key RAM, and retrieved in any order depending upon whether encryption or decryption is in progress. The encryption module processes the plaintext and generates the ciphertext, which is subsequently fed to the decryption module to obtain the plaintext back. The output of the decryption module is compared to a pre-stored copy of the plaintext, and any mismatch results in an error signal and subsequent suppression of the faulty

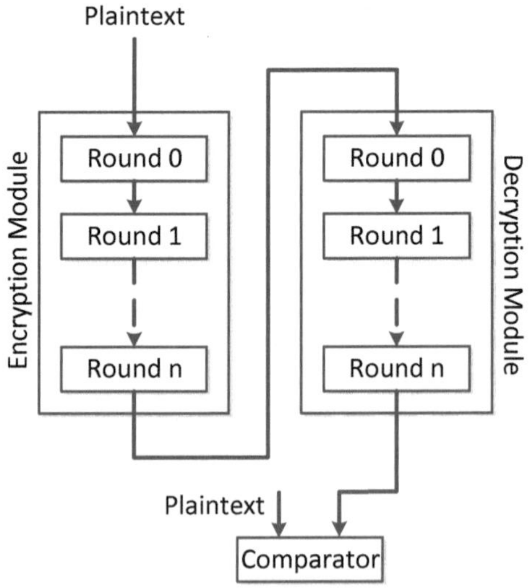

Fig. 8.7 Hybrid redundancy: algorithmic countermeasure

ciphertext. During the decryption operation, the encryption module can be used to process the next batch of plaintext data, thus avoiding significant throughput loss.

8.5.2 *Round-Level Hybrid Redundancy*

The approach described above can be equivalently adopted at the round level, where each round is followed by the corresponding inverted round, and the input of the former is compared to the output of the latter. This approach was applied to AES in [91], and is illustrated in Fig. 8.8. Recall that in AES, the first round of encryption is the inverse of the last round of decryption and so on. The idea is therefore to run encryption and decryption in parallel, with the rounds of the decryption module operated in reverse order, and the output of each encryption round fed to the corresponding inverse round in the decryption module. A comparison of the input to the encryption round with the output of the decryption round is used to detect the presence of any fault.

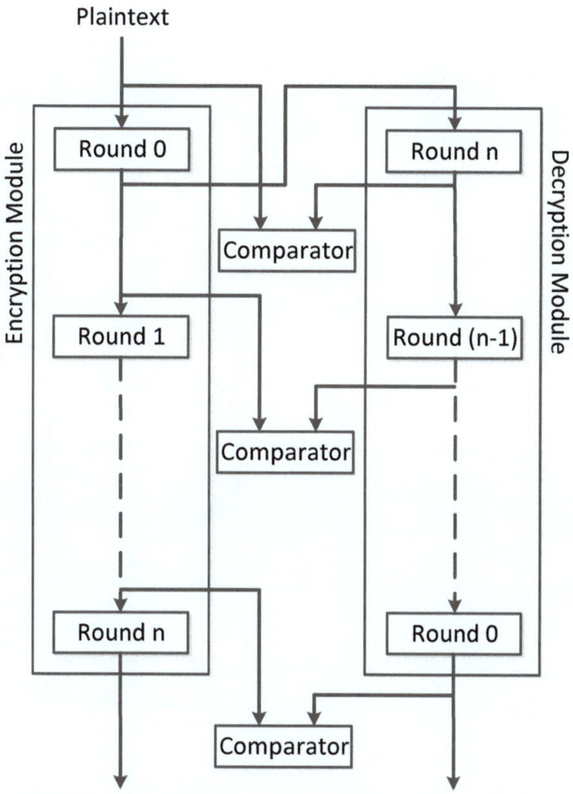

Fig. 8.8 Hybrid redundancy: round-level countermeasure

8.5.3 Round-Level Hybrid Redundancy

An even more granular approach is to break down each encryption round corresponding to a cryptographic algorithm into individual operations, and make each operation redundant using a module for the corresponding inverse operation. Consider, for example, a pipelined implementation of a block cipher, where each round is spread across multiple clock cycles, and each clock cycle processing a single round operation. Figure 8.9 illustrates how such an implementation can be protected using the inverse relationship between the corresponding operations in encryption and decryption rounds. For example, the first operation in the first encryption round is the inverse of the last operation of the last decryption round. Consequently, the output of the former can be fed to the input of the latter, and the corresponding input to the former can be subsequently compared with the output of the latter to detect the presence of faults, if any.

Each of the above countermeasures can be implemented in two modes offering varying area-throughput trade-offs. The first mode is the *half-duplex* mode, where an operation and its inverse are idle when the other counterpart is running. In this mode, the throughput is close to 73.5% of the unprotected computation, however, the same chip can be used for both the original and inverse operations. The second

Fig. 8.9 Hybrid redundancy: operation-level countermeasure

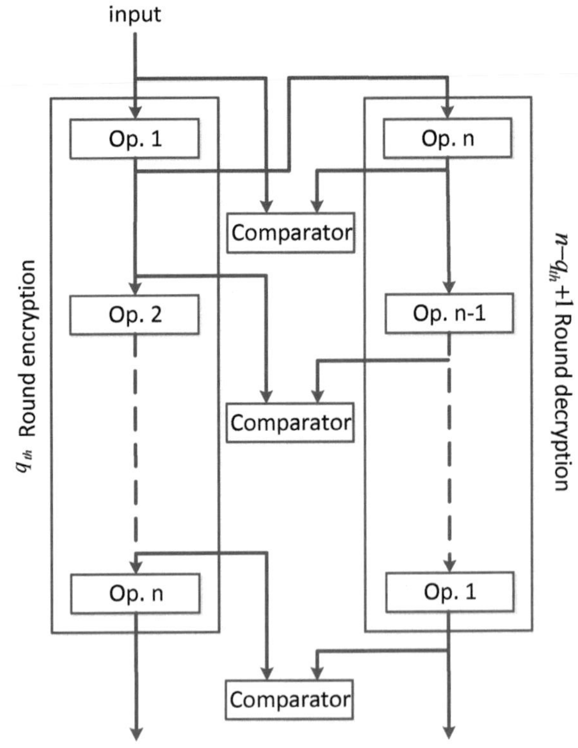

mode is the *full-duplex* mode, where the operation and its inverse module operate in parallel without being idle. In this mode, the throughput is improved to close to 100%; however, it requires separate chips running the original and inversion operations. In some cases, both modules may be present individually in the same chip, thereby reducing the area overhead incurred by the full-duplex mode of operation.

8.5.4 *Optimized Techniques*

An area-optimized version of the countermeasure technique described in Fig. 8.9 was presented by Satoh et al. in [162]. The pipelined architecture for their technique is depicted in Fig. 8.10. Their approach significantly reduces the area requirement by dividing each round function into two sub-blocks, that are used alternately in different stages of the pipeline for original and inverted round operation computations. This avoids the need for any additional error detection block. The only additional requirement is a pipeline register, a selector and a comparator. Although the approach doubles the required number of clock cycles, the maximum frequency is boosted by halving the critical path for each round function via pipelining. The hardware over-

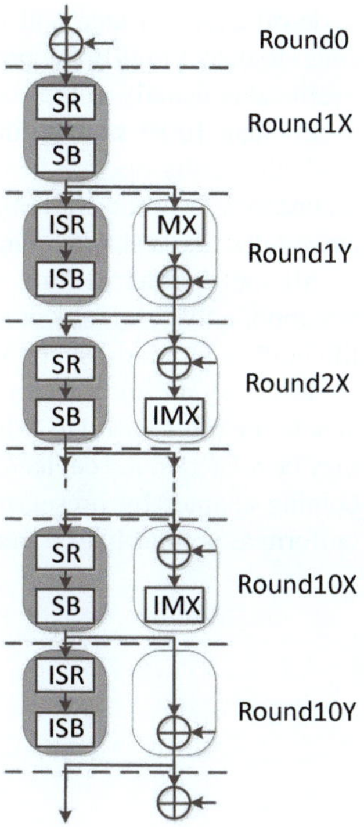

Fig. 8.10 Hybrid redundancy: area-optimized operation-level countermeasure

head for this technique is around 88.9%, as compared to 73.5% in the half-duplex mode of the previous approach, while the hardware overhead is 88.9%, as compared to close to 100% in the full-duplex mode of the previous approach.

8.6 Summary

In this chapter, we have presented to the readers a collection of detection-based countermeasure techniques against DFA. All the countermeasures presented in this chapter use some form of redundancy to detect the presence of a fault, and thereby render the faulty ciphertext unexploitable to the adversary. The techniques can be broadly separated into four categories: temporal redundancy, spatial redundancy, information redundancy, and hybrid redundancy. In temporal redundancy, one or more operations are repeated twice, and the corresponding outputs are compared to detect the presence of a fault. Such techniques usually do not blow up the area significantly, but reduce the overall throughput by nearly half. In spatial redundancy, on the other hand, each operation is performed twice in parallel using dedicated hardware, and the corresponding outputs are compared to detect the presence of a fault. Such techniques usually do not degrade the throughput, but suffer from an area overhead of close to 100%. Information redundancy techniques try and bridge this trade-off between area and throughput by using certain error-detection/correction codes to detect/nullify the presence of a fault in the cipher state. The only additional overhead is usually in the form of prediction units for error-detection, which leads to less than 100% blowup in area, and does not degrade the throughput, since the prediction units operate in parallel to the main encryption block. Finally, hybrid redundancy techniques attempt to combine the benefits of each of the aforementioned techniques into a single countermeasure.

All the countermeasure techniques discussed above are designed under the assumption that all faults in a given fault space occur with equal probability. In real-life implementations, such instances of fault attacks are usually rare to find, wherein an adversary an inject faults uniformly at random. Rather, practical fault injection attacks are often found to exhibit a characteristic bias in the fault-distribution, which may be related to the device/design specifications. We address this issue in the forthcoming chapter by presenting to the readers a generic technique to thwart both uniform as well as biased fault attacks.

Chapter 9
Fault Space Transformation: Countering Biased Fault Attacks

Sikhar Patranabis, Abhishek Chakraborty, Debdeep Mukhopadhyay and P. P. Chakrabarti

Abstract In the previous chapter, we have introduced to the readers several examples of countermeasures against differential fault analysis. These countermeasures are all redundancy-based, and use a concurrent error detection mechanism to infer the presence of a fault. It is important to note here that these classical redundancy-based countermeasures are designed under the assumption that all faults in a given fault space occur with equal probability. In real-life implementations, such instances of fault attacks are usually rare to find, wherein an adversary an inject faults uniformly at random. Rather, practical fault injection attacks are often found to exhibit a characteristic bias in the fault-distribution, which may be related to the device/design specifications. In this chapter, we demonstrate to the readers how fault bias acts as a threat to the security of classical countermeasures.

9.1 Introduction

The previous chapter has introduced to the readers several examples of countermeasures against differential fault analysis (DFA) based on redundancy. Any form of redundancy works under the primary assumption that all potential faults that could disrupt the proper functioning of a given circuit occur with uniformly equal probability. More mathematically, let $\mathscr{F}$ denote the *fault space* - that is, the whole set of possible faults $\{f_1, f_2, \cdots, f_n\}$ that an adversary can inject using a specific fault injection technique. For example, if one were to consider the simplest fault attack on AES-128 presented in [71], where the adversary wishes to recover a single byte of the key, the fault space $\mathscr{F}$ is the set of all possible byte faults $\{1, 2, \cdots, 255\}$. Now, let F be a discrete random variable that denotes the outcome of a single random fault

S. Patranabis (✉) · A. Chakraborty · D. Mukhopadhyay · P. P. Chakrabarti
Secured Embedded Architecture Lab, IIT Kharagpur, Kharagpur, India
e-mail: sikhar.patranabis@iitkgp.ac.in

S. Patranabis and D. Mukhopadhyay (eds.), *Fault Tolerant Architectures for Cryptography and Hardware Security*, Computer Architecture and Design Methodologies, https://doi.org/10.1007/978-981-10-1387-4_9

injection, and let p_i be the probability of occurrence of fault f_i, where $1 \leq i \leq n$. Let $\mathscr{P}$ denote the probability distribution $\{p_1, \cdots, p_n\}$ and refer to it as the fault probability distribution followed by the fault space $\mathscr{F}$. In a uniform fault model, as assumed by most traditional redundancy-based countermeasure strategies, for any pair of faults f_i and f_j, we have $p_i = p_j$, that is, they occur with the same probability. However, in most real-life fault injection scenarios, such uniform fault models are not encountered. This motivates the study of biased fault models, and whether existing countermeasure strategies are equipped well enough to counter such fault models.

9.2 Fault Bias Versus Classical Redundancy

9.2.1 Quantifying the Fault Bias

We attempt to provide a quantitative estimate of the bias of a given fault model based on experimentally observed distribution of fault values. More concretely, given a fault space $\mathscr{F} = \{f_1, f_2, \cdots, f_n\}$, where each fault f_i occurs with probability p_i, we propose a statistical estimator for the bias of the fault probability distribution $\mathscr{P} = \{p_1, \cdots, p_n\}$. Note that for a uniform fault model, all these probabilities are equal; consequently, the metric should ideally capture how far the individual probabilities stray from the uniform distribution. Hence, the variance Var of $\mathscr{P}$, computed as $Var = \frac{\sum_{i=1}^{n} {p_i}^2}{n} - \frac{1}{n^2}$, is ideally suited to be the statistical measure of bias.

9.2.2 Threat to Classical Redundancy

With the aforementioned quantification in place, we investigate the impact of fault bias on the potency of a classical detection-based countermeasure strategy. The adversarial strategy here is straightforward - he tries to inject identical faults in both the original and redundant operations, thereby bypassing the detection mechanism. In most countermeasure strategies, particularly in the case of spatial and redundancy-based countermeasures, the round operations/sub-operations of a block cipher are simply repeated in time and space, respectively; hence it is reasonable to assume that the fault probability distributions during injection operations in both the original and redundant operations is identical. Consequently, the probability of successfully bypassing of the countermeasure is proportional to the probability that the same fault $f_i \in \mathscr{F}$ can be injected in both the original and redundant computation rounds. We refer to this probability as the *fault collision probability*. Let $\hat{f}_0$ and $\hat{f}_1$ be the random variables denoting the outcome of fault injections in the original and redundant rounds respectively. Since the fault injection in the original and redundant rounds are independent, we have $Pr[\hat{f_0} = f_i, \hat{f_1} = f_j] = p_i p_j$. It is straightforward to see that the fault collision probability is given by $\tilde{p} = \sum_{i=1}^{n} Pr[\hat{f_0} = f_i, \hat{f_1} = f_i] = \sum_{i=1}^{n} {p_i}^2$. Interestingly, this leads to the following relation between the fault bias, measured by the variance Var of the fault probability distribution, and the fault collision probability $\tilde{p}$:

$$\tilde{p} = nVar + \frac{1}{n} \tag{9.1}$$

This clearly indicates that for a given fault model, the fault collision probability $\tilde{p}$ is directly proportional to the bias of the underlying fault model. We therefore have a mathematical basis for the following claim: The resistance of a classical redundancy-based countermeasure strategy degrades progressively with the bias of a fault model.

9.3 A Case Study: Biased Fault Attack on Redundant AES-128

In this section, we present an actual experimental study conducted on two different implementations of AES-128 - the first uses temporal redundancy while the latter uses spatial redundancy to counter fault attacks. The fault model is a single byte fault, where any number of bits ranging from one to four are flipped within a given byte of the AES state at the beginning of the last round. Table 9.1 summarizes the abbreviations for the different categories of faults considered in our case study based on the number of flipped bits in the target byte of the AES state. The target platform for our case study is a Spartan-3A XC3S400A FPGA, on to which faults are injected using clock glitches. The device under test operates under two different clock frequencies - a normal frequency for fault-free operation, and a faster clock which is used to induce the glitches the obtain the desired faults. As depicted in Table 9.2(a) and (b), the fault nature varies depending on the glitch frequency for both target implemen-

Table 9.1 Fault model description

Symbol	Fault model
FF	Fault free
SBU	Single bit upset
SBDBU	Single byte double bit upset
SBTBU	Single byte triple bit upset
SBQBU	Single byte quadruple bit upset
OSB	Other single byte faults
MB	Multiple byte faults

tations of AES. Table 9.3 elucidates the common frequency ranges between either round where each type of fault model is predominant for either countermeasure. Note that the monitoring of the fault pattern was performed using a Chipscope pro 12.3 Analyzer, which limits the number of observable samples at a given frequency to 1024. Hence, we observed 512 samples for the original computation and 512 samples for the redundant computation.

Table 9.2 Fault distribution

(a) Fault distribution - temporal redundancy

Fast clock frequency (MHz)	FF	SBU	SBDBU	SBTBU	SBQBU	OSB	MB
125.0	512	0	0	0	0	0	0
125.1	503	9	0	0	0	0	0
125.2	489	22	1	0	0	0	0
125.3	456	50	6	0	0	0	0
125.4	425	59	22	6	0	0	0
125.5	396	45	43	28	0	0	0
125.6	354	34	112	32	0	0	0
125.7	303	23	101	85	0	0	0
125.8	260	11	55	86	0	0	0
125.9	208	5	46	147	6	0	0
126.0	176	1	39	228	68	0	0
126.1	143	0	18	211	136	4	0
126.2	115	0	10	94	178	15	0
126.3	101	0	8	95	251	49	8
126.4	65	0	9	45	232	141	20
126.5	32	0	5	16	131	187	141
126.6	13	0	3	8	98	101	289
126.7	5	0	1	4	32	112	358
126.8	0	0	1	2	5	105	399
126.9	0	0	1	2	3	88	421
127.0	0	0	0	1	2	33	476
127.1	0	0	0	0	1	12	499
127.2	0	0	0	0	0	0	512
127.3	0	0	0	0	0	0	512
127.4	0	0	0	0	0	0	512
127.5	0	0	0	0	0	0	512

(continued)

Table 9.2 (continued)

(b) Fault distribution - spatial redundancy

Fast clock frequency (MHz)	FF	SBU	SBDBU	SBTBU	SBQBU	OSB	MB
70.0	512	0	0	0	0	0	0
70.1	512	0	0	0	0	0	0
70.2	504	8	0	0	0	0	0
70.3	475	34	3	0	0	0	0
70.4	460	47	5	0	0	0	0
70.5	416	63	29	4	0	0	0
70.6	378	38	71	25	0	0	0
70.7	345	29	120	32	0	0	0
70.8	299	21	164	28	0	0	0
70.9	234	14	120	144	2	0	0
71.0	216	4	39	247	6	0	0
71.1	189	2	35	220	66	0	0
71.2	130	0	15	180	176	11	0
71.3	105	0	10	104	278	15	0
71.4	83	0	10	66	227	100	26
71.5	50	0	8	46	157	162	90
71.6	27	0	5	16	113	125	226
71.7	21	0	4	10	98	118	261
71.8	13	0	3	6	50	103	337
71.9	7	0	3	5	21	107	369
72.0	5	0	3	2	10	99	393
72.1	2	0	1	1	8	44	456
72.2	1	0	0	1	6	19	485
72.3	1	0	0	0	2	8	501
72.4	0	0	0	0	1	5	506
72.5	0	0	0	0	0	0	512

Table 9.3 Useful fault models and corresponding frequency ranges

Fault model	Frequency range (MHz)	
	Time redundancy	Hardware redundancy
SBU	125.3–125.4	70.3–70.5
SBDBU	125.6–125.7	70.6–70.9
SBTBU	126.0–126.1	71.0–71.1
SBQBU	126.3–126.4	71.3–71.4

Table 9.4 Experimental results: biased fault attacks on time redundancy

Round	Fault model	Fault variance	Ciphertexts	Faults (estimate)	Faults (practical)
8	SBU	9.5×10^{-2}	305	340	388
	SBDBU	1.4×10^{-2}	625	1456	1448
	SBTBU	9.7×10^{-3}	1020	1816	1975
	SBQBU	3.2×10^{-3}	1879	7869	8003
9	SBU	9.2×10^{-2}	304	386	388
	SBDBU	8.8×10^{-2}	625	641	648
	SBTBU	8.1×10^{-2}	832	874	856
	SBQBU	7.5×10^{-2}	1328	1788	1809

Table 9.5 Experimental results: biased fault attacks on hardware redundancy

Round	Fault model	Fault variance	Ciphertexts	Faults (estimate)	Faults (practical)
8	SBU	1.1×10^{-1}	300	336	323
	SBDBU	9.4×10^{-2}	651	1426	1455
	SBTBU	5.6×10^{-3}	990	1857	1824
	SBQBU	4.5×10^{-3}	1724	7536	7503
9	SBU	9.5×10^{-2}	304	390	377
	SBDBU	7.7×10^{-2}	619	647	664
	SBTBU	7.6×10^{-2}	883	892	829
	SBQBU	3.4×10^{-2}	1299	1851	1913

The next step is to test the required number of fault injections to recover the entire AES-128 key. The attack strategy described in Sect. 9.2.2 was applied on both the time and hardware redundancy countermeasure schemes, using the fault injection setup described above. Tables 9.4 and 9.5 summarize the results of our attacks for different fault models. The estimated number of fault injections required is computed from the knowledge of the fault space and the experimentally observed fault probability variance. The results clearly demonstrate the effectiveness of biased fault attacks in defeating the redundancy-based countermeasures, as compared to uniformly distributed fault models.

9.4 Countering Fault Collision: Transformation of the Fault Space

Given the aforementioned vulnerability of traditional countermeasures to biased fault attacks, the need for a new countermeasure strategy emerges, Such a countermeasure strategy should minimize the average fault collision probability irrespective of the nature of the fault model chosen by the adversary. We introduce to the readers *fault space transformation* [141] - a strategy to minimize fault collision attacks on classical redundancy-based countermeasures. The basic idea here is to perform the original and redundant computations using different encodings, with an equivalence relation W between the respective cipher states after the computations. As a result of this transformation under W, the *equivalent fault space* - the sets of faults to be injected in the original and redundant computation to bypass the countermeasure technique are no longer identical. We explain the approach formally in the following section.

9.4.1 Fault Space Transformation (FST)

Let R_0 and R_1 be the cipher state registers corresponding to the original and redundant computations, respectively, in any redundancy-based countermeasure technique. Let $W : (0, 1)^N \rightarrow (0, 1)^N$ be an additional bijective mapping introduced in the redundant computation such that $R_1 = W(R_0)$ during fault-free operation. Suppose the adversary injects faults in R_0 under a fault space $\mathcal{F}_0$. The equivalent fault space in case of the redundant computation would be some $\mathcal{F}_1$ related to $\mathcal{F}_0$ via the transformation W. In particular, for each fault $f_i \in \mathcal{F}_0$ there is an *equivalent fault* $f_j \in \mathcal{F}_1$, that is, $f_i \equiv f_j$ under the mapping W. Interestingly, $\mathcal{F}_1$ in turn could be a subspace of a much larger fault space $\mathcal{F}^*$. For instance, with respect to AES-128, suppose the mapping W is such that it maps the space of all single byte faults $\mathcal{F}_0$ to a equivalent subspace $\mathcal{F}_1$ comprising entirely of four byte faults (we will demonstrate later that such mappings indeed exist). Quite evidently, $\mathcal{F}_1$ is now a subspace of a much larger fault space $\mathcal{F}^*$, comprising of *all* four byte faults.

The idea behind the countermeasure strategy is that although the adversary may be able to inject biased faults under $\mathcal{F}^*$, the probability that these faults are precisely the faults in $\mathcal{F}_1$ is uniform (and hence low) over all possible choices of W. This may be mathematically explained as follows. Any chosen mapping W maps $\mathcal{F}_0$ to a subset $\mathcal{F}_1$ of $\mathcal{F}^*$ such that $|\mathcal{F}_1| = |\mathcal{F}_0|$. There are $\binom{|\mathcal{F}^*|}{|\mathcal{F}_1|}$ such subsets. A particular fault $f_j \in \mathcal{F}^*$ occurs in $\binom{|\mathcal{F}^*|-1}{|\mathcal{F}_1|-1}$ of the subsets. Thus, given a random fault $f_i \in \mathcal{F}_0$ and a random fault $f_j \in \mathcal{F}^*$, the expectation of $Pr[f_j = W(f_i)]$ over all possible choices of W (assuming the adversary has no control over W) is given as follows:

$$\mathbb{E}(Pr[f_j = W(f_i)]) = \frac{\binom{|\mathcal{F}^*|-1}{|\mathcal{F}_1|-1}}{\binom{|\mathcal{F}^*|}{|\mathcal{F}_1|}|\mathcal{F}_1|} = \frac{1}{|\mathcal{F}^*|} \tag{9.2}$$

Let p_i and p_j be the probability of occurrence of the faults $f_i \in \mathcal{F}_0$ and $f_j \in \mathcal{F}^*$, and let ρ denote the correlation coefficient between the fault probability distributions for $\mathcal{F}_0$ and $\mathcal{F}^*$. Also, let Var_0 and Var^* be the variances of the two fault probability distributions. Assuming that the adversary has perfect knowledge of first fault injection $\hat{f_0} = f_i \in \mathscr{F}_0$ for some i, the expected probability of equivalent fault injection $\tilde{p}$ on two random fault injections $\hat{f_0}$ and $\hat{f_1}$ is given as follows:

$$\begin{aligned}
\mathbb{E}(\tilde{p}) &= \mathbb{E}\left(\sum_{i=1}^{|\mathscr{F}_0|} Pr[\hat{f_0} = f_i, \hat{f_1} = W(f_i)]\right) \\
&= \sum_{i=1}^{|\mathscr{F}_0|} \mathbb{E}(Pr[\hat{f_0} = f_i, \hat{f_1} = W(f_i)]) \\
&= \sum_{i=1}^{|\mathscr{F}_0|} \mathbb{E}(Pr[\hat{f_0} = f_i]Pr[\hat{f_1} = W(f_i)]) \\
&= \sum_{i=1}^{|\mathscr{F}_0|} \mathbb{E}(Pr[\hat{f_0} = f_i])\mathbb{E}(Pr[\hat{f_1} = W(f_i)]) + \rho\sqrt{Var_0.Var^*} \\
&= \sum_{i=1}^{|\mathscr{F}_0|} \mathbb{E}(Pr[\hat{f_0} = f_i])(\frac{1}{|\mathscr{F}^*|}) + \rho\sqrt{Var_0.Var^*} \\
&= \frac{1}{|\mathscr{F}^*|}\sum_{i=1}^{|\mathscr{F}_0|} \mathbb{E}(Pr[\hat{f_0} = f_i]) + \rho\sqrt{Var_0.Var^*} \\
&= \frac{1}{|\mathscr{F}^*|}\mathbb{E}\left(\sum_{i=1}^{|\mathscr{F}_0|} Pr[\hat{f_0} = f_i]\right) + \rho\sqrt{Var_0.Var^*} \\
&= \frac{1}{|\mathscr{F}^*|} + \rho\sqrt{Var_0.Var^*}
\end{aligned} \tag{9.3}$$

9.4.2 Choosing the Transformation W

The next question to answer is how to choose the transformation W. From our discussion above, it is clear that the transformation W should ideally ensure that the fault space $\mathscr{F}_0$ is mapped onto a subspace $\mathscr{F}_1$ of a larger fault space $\mathscr{F}^*$. A larger fault space is necessary to ensure that the adversary cannot inject specifically the desired faults in the small subspace $\mathscr{F}_1$ with high probability. A possible strategy is to use a *Maximum Distance Separable* (MDS) matrix [87] for W. Let W be a $m_2 \times m_1$ linear MDS transformation $\mathbb{K}$ from $\mathbb{K}^{m_1}$ to $\mathbb{K}^{m_2}$. A nice property of MDS matrices is that they guarantee a fault space transformation such that the original and redundant fault spaces $\mathscr{F}_0$ and $\mathscr{F}_1$ differ sufficiently in their Hamming weights. We present an example here with respect to AES-128. Suppose that the adversary injects a single byte fault in the state register corresponding to the original computation. By the property of the MDS transformation, the equivalent fault in the redundant computation must affect at least m_2 bytes to bypass the detection mechanism. Then the precision of the transformed fault space $\mathscr{F}_1$ is approximately $\frac{1}{2^{(8m_2-1)}}$ times lower than the original fault space $\mathscr{F}_0$. This makes it difficult for the adversary to create fault collisions with high probability.

9.5 Case Study: Application of Fault Space Transformation on AES-128

We apply our proposed fault space transformation based countermeasure technique to protect AES-128 against biased fault attacks. For the transformation function W, we propose using the Rijndael MixColumn matrix, which is a 4×4 MDS matrix in the finite field $\mathbf{GF}(2^8)$. Multiplying an input vector of four bytes with this matrix produces a vector of four bytes as output, such that each input byte affects all 4 output bytes. Thus a single byte fault in the input affects results in a four byte fault in the output, implying that the transformation takes place from a single byte fault to a four byte fault. Since fault models beyond single byte faults are of little practical significance to the adversary, it is sufficient to consider the diffusion property of MixColumn for countering single byte fault attacks on the time and hardware redundancy countermeasures.

9.5.1 Realization via Efficient Implementations

Since the MixColumn operation is already present in every round operation for AES-128 except the last round, we can transform a naïve redundancy-based implementation of AES into one with fault space transformation by simply repositioning the state register for the redundant computation. This is illustrated with respect to spatial redundancy in Fig. 9.1. Quite clearly, this implementation incurs no additional overhead due to the incorporation of the transformation between the original and redundant computations. At the same time, the implementation now ensures that the equivalent fault space for the original and redundant operations is no longer the same, with the Rjindael MixColumn matrix ensuring the transformation in fault space that would make it difficult for the adversary to inject equivalent faults in both operations.

9.5.2 Implementation Overhead

The implementation overheads for the spatial and temporal versions of the proposed countermeasure technique are presented in Table 9.6. The results reported correspond to a Spartan-3A XC3S400A FPGA.

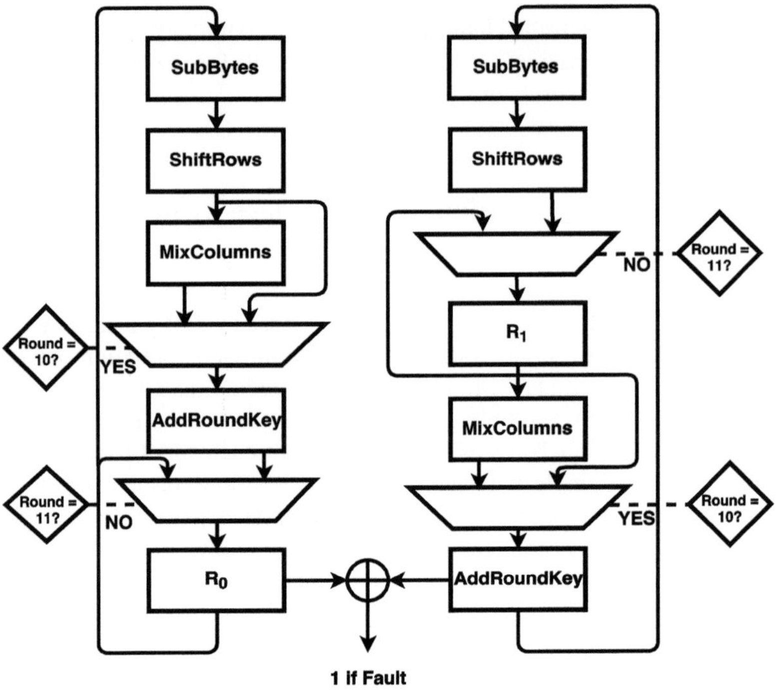

Fig. 9.1 Efficient implementation of AES-128: spatial redundancy with fault space transformation

Table 9.6 Implementation details of our proposed countermeasure

AES implementation	Area overhead (Slice count)	Maximum frequency (MHz)	Clock cycles
Without redundancy	2580	78.5	10
Naïve temporal redundancy	3580	74.6	20
Naïve spatial redundancy	5250	68.9	10
Temporal redundancy + Fault space transformation	3876	72.3	20
Spatial redundancy + Fault space transformation	5538	66.7	10
Optimized architecture (Fig. 9.1)	5356	67.5	10

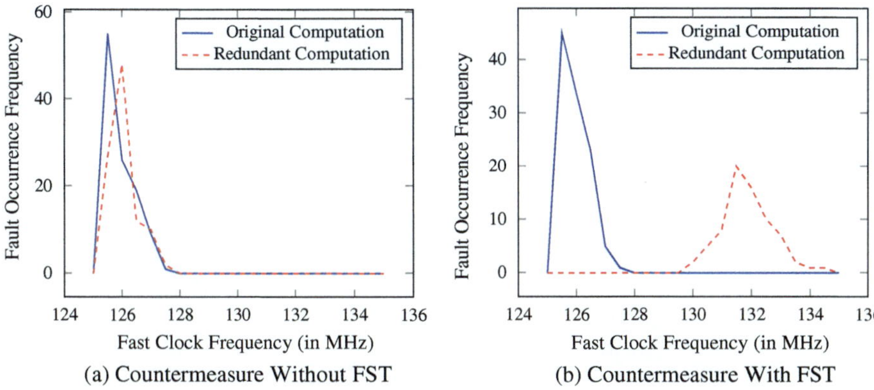

Fig. 9.2 Effect of fault space transformation on biased fault attacks: temporal redundancy

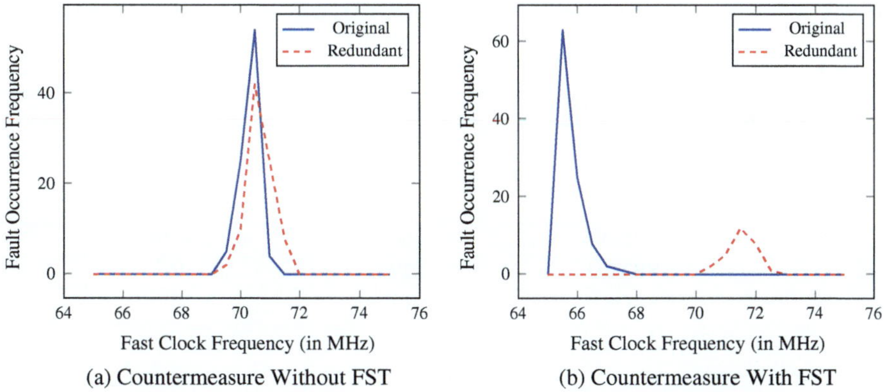

Fig. 9.3 Effect of fault space transformation on biased fault attacks: spatial redundancy

9.5.3 Effect of Fault Space Transformation on Biased Fault Attacks

We finally present to the reader an experimental evaluation of the security of AES-128 with fault space transformation against biased fault attacks. We used a ChipScope pro 12.3 Analyzer to monitor the nature of the faults in the state registers corresponding to the original and redundant computations. As demonstrated in Figs. 9.2 and 9.3, the frequency ranges where the single bit upsets are observed in the original and redundant rounds are completely disjoint due to the presence of the MDS matrix-based transformation between the two computations. Additionally, as indicated in Tables 9.7 and 9.8, the fault collision probability decreases drastically in the presence of FST, thereby ensuring that the number of fault injections necessary to obtain faulty ciphertexts and subsequently recover the entire key by bypassing the countermeasure,

Table 9.7 Effect of fault space transformation with temporal redundancy

Round	Fault model	Faults required (without FST)	Faults required (with FST)
8	SBU	388	3×10^6
	SBDBU	1448	5×10^6
	SBTBU	1975	10^7
	SBQBU	8003	$>10^7$
9	SBU	388	5×10^6
	SBDBU	648	10^7
	SBTBU	856	10^7
	SBQBU	1809	$>10^7$

Table 9.8 Effect of fault space transformation with spatial redundancy

Round	Fault model	Faults required (without FST)	Faults required (with FST)
8	SBU	323	2×10^6
	SBDBU	1455	2.5×10^6
	SBTBU	1824	5×10^6
	SBQBU	7503	$>10^7$
9	SBU	377	3×10^6
	SBDBU	664	5×10^7
	SBTBU	829	$> 10^7$
	SBQBU	1913	$>10^7$

is practically infeasible to obtain. The results hold similarly for each of the fault models considered earlier for the biased fault attack in the absence of FST.

9.6 Summary

In this chapter, we have presented to the readers fault space transformation (FST) - a countermeasure technique against both traditional DFA as well as DFIA-like biased fault attacks on AES-like block ciphers. We presented a formal quantification of the bias of a fault model in terms of the variance of the fault probability distribution, and use this definition to formally argue the threat posed by biased fault attacks to naïve redundancy based countermeasure techniques. We introduced the concept of fault space transformation, in which the adversary is forced to inject two equivalent faults in different fault spaces to bypass the detection step. We argued the use of MDS matrices to provide formal guarantees of low correlation between the original and redundant fault spaces. FST is independent of the block cipher structure, and is generic enough to be applied to a variety of redundancy based countermeasures against a variety of fault injection techniques. We also presented a case study on AES-128 to prove the effectiveness of this countermeasure. We demonstrated how FST can be used to thwart glitch based biased fault attacks on RTL implementations of spatial and temporal redundant implementations of AES-128 on a Spartan 3A FPGA on a SASEBO GII board.

Chapter 10
Infective Countermeasures Against Fault Analysis

Sikhar Patranabis and Debdeep Mukhopadhyay

Abstract The previous discussion has illustrated to the reader the vulnerabilities of classical redundancy based countermeasure techniques, and potential workarounds to avoid the same via fault space transformation. In this chapter, we introduce a different flavor of countermeasures against fault analysis - infective countermeasures. Infective countermeasures are superior to detection based countermeasures in the sense that they avoid the use of explicit comparison steps that are themselves vulnerable to fault attacks. Infective countermeasures can be broadly classified into two categories - deterministic and randomized. Since all deterministic infective countermeasures have been demonstrated to be insecure in principle (Lomné et al, Fault diagnosis and tolerance in Cryptography – FDTC 2012, 2012, [114]), we focus on state-of-the-art randomized infective countermeasures in this chapter. We present to the reader an infective countermeasure for AES-128 proposed by Gierlichs et al. (Progress in cryptology – LATINCRYPT 2012, 2012, [70]), which was the first randomized infective countermeasure to be proposed in the literature. Unfortunately, this countermeasure is found to have certain vulnerabilities against fault attacks (Battistello and Giraud, Fault diagnosis and tolerance in cryptography – FDTC 2013, 2013, [22]), which we subsequently present to the reader. Finally, we present to the reader a second infective countermeasure for AES-128 proposed by Tupsamudre et al. (Cryptographic Hardware and Embedded Systems–CHES 2014, 2014, [173]) that successfully overcomes these vulnerabilities, and is currently the most secure infective countermeasure in the literature.

10.1 Introduction

Infective countermeasures combine a diffusion technique with the original cryptographic operation to detect the presence the presence of a fault. The main difference between an infective countermeasure and a detection-based countermeasure is that

S. Patranabis (✉) · D. Mukhopadhyay
Secured Embedded Architecture Lab, IIT Kharagpur, Kharagpur, India
e-mail: sikhar.patranabis@iitkgp.ac.in

S. Patranabis and D. Mukhopadhyay (eds.), *Fault Tolerant Architectures for Cryptography and Hardware Security*, Computer Architecture and Design Methodologies, https://doi.org/10.1007/978-981-10-1387-4_10

the former avoids the use of an explicit comparison step between the original and redundant operation, which itself could be vulnerable to potential attacks. Instead it works as follows: if an injected fault affects an internal operation of the target algorithm, the diffusion expands and randomizes the internal state of the algorithm so as to render the final erroneous ciphertext unexploitable from an adversary's perspective. On the other hand, in the absence of a fault, the diffusion scheme does not expand itself into the functioning of the encryption algorithm, thereby preserving the correct operational output. In this chapter, we present to the readers two variants of infective countermeasures to protect implementations of AES-128 against DFA attacks. We also present information theoretic guarantees of security for some of these countermeasures, against both single order and higher order fault injections. We then present a third alternative countermeasure strategy that prevents other advanced attacks discussed previously, namely instruction skip attacks, at an algorithmic level.

We introduce some notations used throughout the rest of the chapter. We denote by $RoundFunction$ an entire round of the AES-128 block cipher, comprising of the SubByte, ShiftRow and MixColumn transformations, followed by the AddRoundKey operation. For $i \in [0, 10]$, the ith such $RoundFunction$ takes as input a 16 byte state matrix $\mathbf{I^i}$ and a 16 byte matrix $\mathbf{k^{i+1}}$ that represents ith round key, derived from the master secret-key $\mathbf{K}$. We denote by $\mathbf{S}$ the SubByte operation, by $\mathbf{SR}$ the ShiftRow operation, and by $\mathbf{MC}$ the MixColumn operation within the $RoundFunction$. We also denote by $(m_0 \ldots m_{15})$ the 16 bytes of the AES state matrix, arranged in a 4 $\times$ column-major array.

10.2 The First Infective Countermeasure for AES-128

In this section, we present the first randomized infective countermeasure for AES-128 proposed by Gierlichs et al. in [70]. Algorithm 10 summarizes the infective countermeasure algorithm. The algorithm begins by copying the plaintext P to the registers R_0 and R_1, and a secret value β to the register R_2. Each round of AES is repeated twice - one of these rounds is called the cipher round, while the other is referred to as the redundant round. The redundant round operates on the register R_1, and precedes the cipher round, which operates on the register R_0. The whole algorithm is interspersed by numerous dummy rounds which occur at random time instances, and a compulsory dummy round is executed in Step 14 of the algorithm. The dummy round takes as input a secret value β and an idempotent secret key k^0, chosen such that $RoundFunction(\beta, k^0) = \beta$. The round signatures are hidden carefully from the adversary by making all rounds equivalent with respect to side-channel leakages. This is achieved via the introduction of several dummy operations such as dummy SubByte, ShiftRow and MixColumn operations in the 0th round, and a dummy MixColumn operation in the 10th round of AES-128. The registers C_0, C_1 and C_2 store the intermediate computation results of various cipher, redundant

Algorithm 10 The First Infective Countermeasure

Input: P, k^j for $j \in \{1, \ldots, n\}$, (β, k^0), $(n = 11)$ for AES-128
1:
Output: C = BlockCipher(P, K)
2: State $R_0 \leftarrow P$, Redundant state $R_1 \leftarrow P$, Dummy state $R_2 \leftarrow \beta$
3: $C_0 \leftarrow 0, C_1 \leftarrow 0, C_2 \leftarrow \beta, i \leftarrow 1$
4: **while do**$i \leq 2n$ do:
5: $\lambda \leftarrow RandomBit()$ // $\lambda = 0$ implies a dummy round
6: $\kappa \leftarrow (i \wedge \lambda) \oplus 2(\neg\lambda)$
7: $\zeta \leftarrow \lambda \cdot \lceil i/2 \rceil$ // ζ is actual round counter, 0 for dummy
8: $R_\kappa \leftarrow RoundFunction(R_\kappa, k^\zeta)$
9: $C_\kappa \leftarrow R_\kappa \oplus C_2 \oplus \beta$ // infect C_κ to propagate a fault
10: $\varepsilon \leftarrow \lambda(\neg(i \wedge 1)) \cdot SNLF(C_0 \oplus C_1)$ // check if i is even
11: $R_2 \leftarrow R_2 \oplus \varepsilon$
12: $R_0 \leftarrow R_0 \oplus \varepsilon$
13: $i \leftarrow i + \lambda$
14: **end while**
15: $R_0 \leftarrow R_0 \oplus RoundFunction(R_2, k^0) \oplus \beta$
16: **return** R_0

and dummy rounds, respectively. A random bit λ decides the execution path of the countermeasure algorithm as follows:

1. $\lambda = 0$, dummy round is executed.
2. $\lambda = 1$ and parity of i is even, cipher round is executed.
3. $\lambda = 1$ and parity of i is odd, redundant round is executed.

The computation of a cipher round is followed by an equality check of C_0 and C_1 (which is currently expected to store the output of the corresponding redundant round). The XOR differential of these registers is transformed under a byte-wise non-linear transformation, referred to as Some Non-Linear Function ($SNLF$). The transformation maps a zero bye to zero, and all other non-zero input bytes to non-zero values. The infection created by this transformation spreads across the entire computation under the following circumstances:

- In case of a fault injection in either a cipher or a redundant round, in which case the differential of C_0 and C_1 is non-zero (steps 9–11 of Algorithm 10).
- In case of a fault injection in a dummy round, in which case $C_2 \neq \beta$(steps 8–11 of Algorithm 10).

Finally, Step 14 XOR-s the output of the last cipher round with the output of dummy round and β. The resulting value is returned by the algorithm in Step 15.

The randomized nature of the dummy round operations implies that an adversary can never be certain regarding the nature of the targeted round. For example, in an unprotected implementation of AES-128, a fault injected at the inout of the 9th round spreads to exactly four bytes of the output ciphertext. On the other hand, in the infective countermeasure, described above the presence of compulsory dummy rounds ensures that the fault diffuses to all bytes of the ciphertext. Hence, the adversary cannot ascertain if the fault was injected in a cipher round, redundant round or a dummy round. In order to bypass the injection step, the adversary must perform a double fault injection in a pair of cipher and redundant rounds. However, since the dummy rounds are distributed uniformly at random across the entire execution of the algorithm, the adversary cannot ascertain when exactly to inject the fault twice.

10.2.1 Some Loopholes in the Infective Countermeasure

Despite its apparent infallibility against DFA, the infective countermeasure presented above has been found to contain certain loopholes which may be exploited to recover the secret key. The first such loophole was pointed out by Battistello et al. [22], who showed how to attack the infective countermeasure using a *constant byte fault model*. They observed that only one dummy round occurs after the 10th cipher round in Algorithm 10. This in turn implies that even if the computation of the 10th round is found to be faulty, the infection is limited to only 4 bytes of the output, instead of the whole stat matrix as is ideally desirable. This makes it possible to launch an attack as described next. Assume that the adversary injects a fault f in I_1^{10}, i.e. the first byte of second row in 10th cipher round input I^{10}. The differential between the cipher round and redundant round outputs, computed in Step 7 of Algorithm 10 is:

$$R_0 \oplus R_1 = \begin{pmatrix} 0 & 0 & 0 & 0 \\ 0 & 0 & 0 & \varepsilon \\ 0 & 0 & 0 & 0 \\ 0 & 0 & 0 & 0 \end{pmatrix}$$

where $\varepsilon = S[I_1^{10} \oplus f] \oplus S[I_1^{10}]$. Subsequently, the infection affects the registers R_2 and R_0 in steps 10 and 11 of Algorithm 10, respectively, leading to the following differential:

$$R_0 \oplus R_1 = \begin{pmatrix} 0 & 0 & 0 & 0 \\ 0 & 0 & 0 & \varepsilon \oplus SNLF[\varepsilon] \\ 0 & 0 & 0 & 0 \\ 0 & 0 & 0 & 0 \end{pmatrix}$$

In the final step (Step 14) of Algorithm 10, the dummy round operates on an already infected register R_2, which in turn again infects R_0. However, the presence of the ShiftRow operation shifts the infection entirely to column 3, leaving the faulty byte of R_0 in column 4 unmasked. The overall differential between the correct ciphertext C and faulty C^* may be summarized as:

$$\therefore C \oplus C^* = \begin{pmatrix} 0 & 0 & \beta_8' \oplus \beta_8 & 0 \\ 0 & 0 & \beta_9' \oplus \beta_9 & \varepsilon \oplus SNLF[\varepsilon] \\ 0 & 0 & \beta_{10}' \oplus \beta_{10} & 0 \\ 0 & 0 & \beta_{11}' \oplus \beta_{11} & 0 \end{pmatrix} \tag{10.1}$$

where $\beta'_8, \beta'_9, \beta'_{10}, \beta'_{11}$ are the infected bytes of the compulsory dummy round output. Now, one can easily retrieve the key byte k^{11}_{13} of the last round key from the ciphertext bye C^*_{13}, which is entirely unaffected by the infected output of the dummy round. One can repeat the procedure similarly to retrieve a further 11 bytes of the last round key.

Note that the aforementioned attack procedure only targets those bytes of the faulty intermediate state that are unmasked by the infection spreading from the erroneous dummy round. It may seem that if the infection spreads to every single byte of the 10th round output, the aforementioned attack [22] can be thwarted. Unfortunately, this is not the case. Tupsamudre et al. [173] subsequently proposed an attack which demonstrates that the infection caused by the dummy round is deterministic and can hence be removed. The attack is described next.

10.2.2 Infection Caused by a Dummy Round

Note that in Algorithm 10, a dummy round has the same input and output, namely β. In other words, we have:

$$MC(SR(S(\beta))) \oplus k^0 = \beta$$

which in turn leads to the following expression for Step 14 of Algorithm 10:

$$\begin{aligned} RoundFunction(R_2, k^0) \oplus \beta &= MC(SR(S(R_2))) \oplus k^0 \oplus MC(SR(S(\beta))) \oplus k^0 \\ &= MC(SR(S(R_2))) \oplus MC(SR(S(\beta))) \end{aligned}$$

which may further be represented as:

$$\therefore RoundFunction(R_2, k^0) \oplus \beta = MC(SR(S(R_2) \oplus S(\beta))) \tag{10.2}$$

It is now clear that even if a fault is detected during the compulsory dummy round, that is, $R_2 \neq \beta$, the resulting infection on the final output from the register R_0 can be removed using the aforementioned equation. This leaks the faulty ciphertext to the adversary. Based on Eq. 10.2, we have the following relation between the correct ciphertext C and the faulty ciphertext C^*:

$$C \oplus C^* = \begin{pmatrix} 0 & 0 & 3 \cdot x & 0 \\ 0 & 0 & 2 \cdot x & \varepsilon \oplus SNLF[\varepsilon] \\ 0 & 0 & 1 \cdot x & 0 \\ 0 & 0 & 1 \cdot x & 0 \end{pmatrix}$$

where $x = S[\beta_{13} \oplus SNLF[\varepsilon]] \oplus S[\beta_{13}]$. This forms the basis for the first attack, which exploits the execution of the infected compulsory dummy round.

10.3 Execution of Infected Compulsory Dummy Round

Suppose that a fault f infects the byte β_{13} of the register R_2 after the 10th cipher round has already been executed.

$$R_2 = R_2 \oplus \varepsilon = \begin{pmatrix} \beta_0 & \beta_4 & \beta_8 & \beta_{12} \\ \beta_1 & \beta_5 & \beta_9 & \beta_{13} \oplus SNLF[\varepsilon] \\ \beta_2 & \beta_6 & \beta_{10} & \beta_{14} \\ \beta_3 & \beta_7 & \beta_{11} & \beta_{15} \end{pmatrix}$$

where $\varepsilon = S[I_1^{10}] \oplus S[I_1^{10} \oplus f]$. Now, since the input to the compulsory dummy round is infected, the execution of the dummy round is as demonstrated next. Following the ShiftRow and SubByte operations, the register R_2 takes the form:

$$R_2 = \begin{pmatrix} S[\beta_0] & S[\beta_4] & S[\beta_8] & S[\beta_{12}] \\ S[\beta_5] & S[\beta_9] & S[\beta_{13} \oplus SNLF[\varepsilon] & S[\beta_1] \\ S[\beta_{10}] & S[\beta_{14}] & S[\beta_2] & S[\beta_6] \\ S[\beta_{15}] & S[\beta_3] & S[\beta_7] & S[\beta_{11}] \end{pmatrix}$$

For ease of understanding, we show the output of the MixColumn and AddRoundKey operations only for the third column of the state matrix:

$$\beta_8' = 2 \cdot S[\beta_8] \oplus 3 \cdot S[\beta_{13} \oplus SNLF[\varepsilon]] \oplus 1 \cdot S[\beta_2] \oplus 1 \cdot S[\beta_7] \oplus k_8^0$$
$$\beta_9' = 1 \cdot S[\beta_8] \oplus 2 \cdot S[\beta_{13} \oplus SNLF[\varepsilon]] \oplus 3 \cdot S[\beta_2] \oplus 1 \cdot S[\beta_7] \oplus k_9^0$$
$$\beta_{10}' = 1 \cdot S[\beta_8] \oplus 1 \cdot S[\beta_{13} \oplus SNLF[\varepsilon]] \oplus 2 \cdot S[\beta_2] \oplus 3 \cdot S[\beta_7] \oplus k_{10}^0$$
$$\beta_{11}' = 3 \cdot S[\beta_8] \oplus 1 \cdot S[\beta_{13} \oplus SNLF[\varepsilon]] \oplus 1 \cdot S[\beta_2] \oplus 2 \cdot S[\beta_7] \oplus k_{11}^0$$

Since $RoundFunction(\beta, k^0) = \beta$, we can write the 3rd column of β as:

$$\beta_8 = 2 \cdot S[\beta_8] \oplus 3 \cdot S[\beta_{13}] \oplus 1 \cdot S[\beta_2] \oplus 1 \cdot S[\beta_7] \oplus k_8^0$$
$$\beta_9 = 1 \cdot S[\beta_8] \oplus 2 \cdot S[\beta_{13}] \oplus 3 \cdot S[\beta_2] \oplus 1 \cdot S[\beta_7] \oplus k_9^0$$
$$\beta_{10} = 1 \cdot S[\beta_8] \oplus 1 \cdot S[\beta_{13}] \oplus 2 \cdot S[\beta_2] \oplus 3 \cdot S[\beta_7] \oplus k_{10}^0$$
$$\beta_{11} = 3 \cdot S[\beta_8] \oplus 1 \cdot S[\beta_{13}] \oplus 1 \cdot S[\beta_2] \oplus 2 \cdot S[\beta_7] \oplus k_{11}^0$$

Finally, the output of the compulsory dummy round is XOR-ed with β in Step 14 of Algorithm 10:

$$\therefore RoundFunction(R_2, k^0) \oplus \beta = \begin{pmatrix} 0 & 0 & 3 \cdot S[\beta_{13} \oplus SNLF[\varepsilon]] \oplus 3 \cdot S[\beta_{13}] & 0 \\ 0 & 0 & 2 \cdot S[\beta_{13} \oplus SNLF[\varepsilon]] \oplus 2 \cdot S[\beta_{13}] & 0 \\ 0 & 0 & 1 \cdot S[\beta_{13} \oplus SNLF[\varepsilon]] \oplus 1 \cdot S[\beta_{13}] & 0 \\ 0 & 0 & 1 \cdot S[\beta_{13} \oplus SNLF[\varepsilon]] \oplus 1 \cdot S[\beta_{13}] & 0 \end{pmatrix}$$

From the point of view of the designer, the ideal countermeasure strategy would be to ensure that every byte of the faulty ciphertext C^* is infected via masking with an independent and uniformly random value. Unfortunately, in Algorithm 10, the compulsory dummy round infects only column 3 of C^* with interrelated values, while the remaining bytes are left unmasked.

10.3.1 An Attack on the Top Row

One may now use the above observation to launch a fault attack targeting the topmost row of of I^{10} (the input to the 10th round of AES-128). The attack improves upon the previously stated attack by Battistello et al. in the sense that it is additionally capable of uniquely recovering the remaining four bytes of the last round key. The previous attack recovered these key bytes via a brute force search. Suppose that a fault f is injected in the first byte of the 10th cipher round input I^{10}. We then have the following differential relation between the faulty register R_0 and the redundant state register R_1, in Step 7 of Algorithm 10:

$$R_0 \oplus R_1 = \begin{pmatrix} \varepsilon & 0 & 0 & 0 \\ 0 & 0 & 0 & 0 \\ 0 & 0 & 0 & 0 \\ 0 & 0 & 0 & 0 \end{pmatrix}$$

where $\varepsilon = S[I_0^{10} \oplus f] \oplus S[I_0^{10}]$. Subsequently, the infection affects the registers R_2 and R_0 in steps 10 and 11 of Algorithm 10, respectively, leading to the following differential:

$$R_0 \oplus R_1 = \begin{pmatrix} \varepsilon \oplus SNLF[\varepsilon] & 0 & 0 & 0 \\ 0 & 0 & 0 & 0 \\ 0 & 0 & 0 & 0 \\ 0 & 0 & 0 & 0 \end{pmatrix}$$

In the final step (Step 14) of Algorithm 10, the dummy round operates on an already infected register R_2, which in turn again infects R_0. However, in this case, the ShiftRow operation does not shift the infection, implying that the resultant final difference between the correct ciphertext C and faulty ciphertext C^* is given by:

$$\therefore C \oplus C^* = \begin{pmatrix} \varepsilon \oplus SNLF[\varepsilon] \oplus \beta_0' \oplus \beta_0 & 0 & 0 & 0 \\ \beta_1' \oplus \beta_1 & 0 & 0 & 0 \\ \beta_2' \oplus \beta_2 & 0 & 0 & 0 \\ \beta_3' \oplus \beta_3 & 0 & 0 & 0 \end{pmatrix} \qquad (10.3)$$

where $\beta_0', \beta_1', \beta_2', \beta_3'$ are random masking outputs from the compulsory dummy round, and $\varepsilon = S[I_0^{10} \oplus f] \oplus S[I_0^{10}]$. Clearly, the key-recovery technique of Battistello et al. cannot be applied directly here due to the presence of the masking from the infected dummy round. Hence, in order to replicate the key-recovery attack in this case, we need the following additional effort: we must remove the infection caused by the dummy round. From the previously stated relations, we can write the aforementioned differential matrix as:

$$C \oplus C^* = \begin{pmatrix} \varepsilon \oplus SNLF[\varepsilon] \oplus 2 \cdot y & 0 & 0 & 0 \\ 1 \cdot y & 0 & 0 & 0 \\ 1 \cdot y & 0 & 0 & 0 \\ 3 \cdot y & 0 & 0 & 0 \end{pmatrix} \tag{10.4}$$

where $y = S[\beta_0 \oplus SNLF[\varepsilon]] \oplus S[\beta_0]$. Following the execution of the 10th cipher round, if an adversary injects a fault f in the state register I_0^{10}, it infects the byte β_0 of R_2 in Step 10 of Algorithm 10:

$$R_2 = R_2 \oplus \varepsilon = \begin{pmatrix} \beta_0 \oplus SNLF[\varepsilon] & \beta_4 & \beta_8 & \beta_{12} \\ \beta_1 & \beta_5 & \beta_9 & \beta_{13} \\ \beta_2 & \beta_6 & \beta_{10} & \beta_{14} \\ \beta_3 & \beta_7 & \beta_{11} & \beta_{15} \end{pmatrix}$$

where $\varepsilon = \mathrm{S}[I_0^{10}] \oplus S[I_0^{10} \oplus f]$. The compulsory dummy round is now executed on the infected input R_2 as demonstrated below:

$$R_2 = \begin{pmatrix} S[\beta_0] \oplus SNLF[\varepsilon] & S[\beta_4] & S[\beta_8] & S[\beta_{12}] \\ S[\beta_5] & S[\beta_9] & S[\beta_{13} & S[\beta_1] \\ S[\beta_{10}] & S[\beta_{14}] & S[\beta_2] & S[\beta_6] \\ S[\beta_{15}] & S[\beta_3] & S[\beta_7] & S[\beta_{11}] \end{pmatrix}$$

Once again, for the ease of understanding, we demonstrate the output of MixColumn and AddRoundKey operations with respect to only the 3rd column:

$$\begin{aligned}
\beta_0' &= 2 \cdot S[\beta_0 \oplus SNLF[\varepsilon]] \oplus 3 \cdot S[\beta_5] \oplus 1 \cdot S[\beta_{10}] \oplus 1 \cdot S[\beta_{15}] \oplus k_0^0 \\
\beta_1' &= 1 \cdot S[\beta_0 \oplus SNLF[\varepsilon]] \oplus 2 \cdot S[\beta_5] \oplus 3 \cdot S[\beta_{10}] \oplus 1 \cdot S[\beta_{15}] \oplus k_1^0 \\
\beta_2' &= 1 \cdot S[\beta_0 \oplus SNLF[\varepsilon]] \oplus 1 \cdot S[\beta_5] \oplus 2 \cdot S[\beta_{10}] \oplus 3 \cdot S[\beta_{15}] \oplus k_2^0 \\
\beta_3' &= 3 \cdot S[\beta_0 \oplus SNLF[\varepsilon]] \oplus 1 \cdot S[\beta_5] \oplus 1 \cdot S[\beta_{10}] \oplus 2 \cdot S[\beta_{15}] \oplus k_3^0
\end{aligned}$$

The idempotent nature of the dummy round function allows us to write the following equations using the random masking components:

$$\begin{aligned}
\beta_0' &= 2 \cdot S[\beta_0] \oplus 3 \cdot S[\beta_5] \oplus 1 \cdot S[\beta_{10}] \oplus 1 \cdot S[\beta_{15}] \oplus k_0^0 \\
\beta_1' &= 1 \cdot S[\beta_0] \oplus 2 \cdot S[\beta_5] \oplus 3 \cdot S[\beta_{10}] \oplus 1 \cdot S[\beta_{15}] \oplus k_1^0 \\
\beta_2' &= 1 \cdot S[\beta_0] \oplus 1 \cdot S[\beta_5] \oplus 2 \cdot S[\beta_{10}] \oplus 3 \cdot S[\beta_{15}] \oplus k_2^0 \\
\beta_3' &= 3 * S[\beta_0] \oplus 1 \cdot S[\beta_5] \oplus 1 \cdot S[\beta_{10}] \oplus 2 \cdot S[\beta_{15}] \oplus k_3^0
\end{aligned}$$

As already mentioned, the final step (the compulsory dummy round) involves an XOR of the result of the aforementioned computation with the constant β as:

$$\therefore RoundFunction(R_2, k^0) \oplus \beta = \begin{pmatrix} 2 \cdot S[\beta_0 \oplus SNLF[\varepsilon]] \oplus 2 \cdot S[\beta_0] & 0 & 0 & 0 \\ 1 \cdot S[\beta_0 \oplus SNLF[\varepsilon]] \oplus 1 \cdot S[\beta_0] & 0 & 0 & 0 \\ 1 \cdot S[\beta_0 \oplus SNLF[\varepsilon]] \oplus 1 \cdot S[\beta_0] & 0 & 0 & 0 \\ 3 \cdot S[\beta_0 \oplus SNLF[\varepsilon]] \oplus 3 \cdot S[\beta_0] & 0 & 0 & 0 \end{pmatrix}$$

One can now remove the infection from $C \oplus C^*$ using the knowledge of $1 \cdot y$ to retrieve the erroneous byte from the masked output. This allows launching the same attack as proposed by Battistello et al. to recover k_0^{11}. In other words, the attack of Battistello et al. is now generic enough to target any byte in the state matrix, instead of only the last three rows.

10.4 A Second Attack: Piret and Quisquater's Attack

Tupsamudre et al. also proposed a second, even stronger, fault attack targeting the penultimate round, that is, the 9th of the infective ountermeasure depicted in Algorithm 10. The compulsory dummy round ensures that a fault injected in the 9th cipher round input would infect all bytes of the final output ciphertext. The attack proposed by Tupsamudre et al. demonstrates how, even in this case, the adversary could remove the infection due to the complusory dummy round using the previously obtained relations, and unmask the faulty ciphertext. This then allows launching Piret and Quisquater's fault attack on the faulty ciphertext to recover multiple bytes of the last round key. The attack has the maximum efficiency in the absence of dummy rounds other than the final compulsory dummy round. In such cases, the entire last round key can be recovered using only 8 fault injections. However, the attack is powerful enough to encompass a limited number of randomly occurring dummy rounds.

10.4.1 Basic Attack Procedure

In the absence of any dummy round (except the final compulsory one which is unavoidable, provided that the adversary does not skip instructions or alter the flow of the algorithm via fault injections), the attack procedure involves the following simple steps:

- Observe that if a fault were injected in the input of the 9th cipher round, the cipher state would be infected thrice - once after the 9th cipher round, once after the 10th cipher round, and once after the compulsory dummy round. From the nature of the faulty ciphertext, one needs to analyze and identify if the fault was appropriately injected at the beginning of the 9th cipher round. The analysis process is non-trivial, and is discussed later.

- The next step is to remove the effect of the infections caused by the compulsory dummy round, and also the 10th and subsequently, the 9th cipher round.
- Subsequently, one can apply the standard attack procedure of Piret and Quisquater [147] to recover the last round key.

10.4.2 Identifying Desired Faulty Ciphertexts

As already mentioned, an important step of the attack is to filter out the faulty ciphertexts corresponding to appropriate fault injections in the input of the 9th round. For Algorithm 10, one can verify if a fault was injected in the ith column of the AES state matrix by checking the validity of the following relations between the correct ciphertext T and the faulty ciphertext T^*:

$$\begin{aligned}
(T \oplus T^*)_{(4\cdot(i+1))\%16} &= (T \oplus T^*)_{(4\cdot(i+1))\%16+1} \\
(T \oplus T^*)_{(4\cdot(i+1))\%16+2} &= 3 \cdot (T \oplus T^*)_{(4\cdot(i+1))\%16} \\
(T \oplus T^*)_{(4\cdot(i+3))\%16+2} &= (T \oplus T^*)_{(4\cdot(i+3))\%16+3} \\
(T \oplus T^*)_{(4\cdot(i+3))\%16} &= 3 \cdot (T \oplus T^*)_{(4\cdot(i+3))\%16+2}
\end{aligned} \tag{10.5}$$

where $(T \oplus T^*)_j$ represents the jth byte in the differential $T \oplus T^*$. Since the compulsory dummy round uses the same mask to infect multiple bytes in the AES state matrix, the aforementioned relation allow inference of the fault injection pattern from an adversary's point of view.

10.4.3 Overview of the Attack Procedure

We now provide the readers with an overview of the attack procedure, under the assumption that the only dummy round to occur is the compulsory dummy round, and a fault is successfully injected in the first byte of the 9th cipher round input. Post the execution of the compulsory dummy round, the following differential relations are observed between the correct and faulty ciphertexts T and T^*, respectively:

$$T \oplus T^* = \begin{pmatrix} m_0 \oplus 2F_1 \oplus 1F_2 & 1F_3 & 3F_4 \oplus 1F_5 \oplus 1F_6 & 3F_7 \\ 1F_1 \oplus 3F_2 & 1F_3 & 2F_4 \oplus 3F_5 \oplus 1F_6 & m_1 \oplus 2F_7 \\ 1F_1 \oplus 2F_2 & 3F_3 & m_2 \oplus 1F_4 \oplus 2F_5 \oplus 3F_6 & 1F_7 \\ 3F_1 \oplus 1F_2 & m_3 \oplus 2F_3 & 1F_4 \oplus 1F_5 \oplus 2F_6 & 1F_7 \end{pmatrix} \tag{10.6}$$

where $F_i, i \in \{1, \ldots, 7\}$, summarize the infection caused by the compulsory dummy round in Step 14 and m_j, $j \in \{0, 1, 2, 3\}$, represent the differential between the correct and faulty cipher states prior to the final infection from the dummy round. Let us now take a step back and analyze the differential between the faulty and redundant registers R_0 and R_1, respectively, after the execution of the 9th cipher round:

$$R_0 \oplus R_1 = \begin{pmatrix} A & 0 & 0 & 0 \\ B & 0 & 0 & 0 \\ C & 0 & 0 & 0 \\ D & 0 & 0 & 0 \end{pmatrix}$$

where $A = 2 * f'$, $B = 1 * f'$, $C = 1 * f'$ and $D = 3 * f'$. This is followed by the first infection, which results in the following differential between the faulty and redundant registers:

$$R_0 \oplus R_1 = \begin{pmatrix} A \oplus SNLF[A] & 0 & 0 & 0 \\ B \oplus SNLF[B] & 0 & 0 & 0 \\ C \oplus SNLF[C] & 0 & 0 & 0 \\ D \oplus SNLF[D] & 0 & 0 & 0 \end{pmatrix}$$

Additionally, the infection also spreads to the dummy register R_2:

$$R_2 = \begin{pmatrix} \beta_0 \oplus SNLF[A] & \beta_4 & \beta_8 & \beta_{12} \\ \beta_1 \oplus SNLF[B] & \beta_5 & \beta_9 & \beta_{13} \\ \beta_2 \oplus SNLF[C] & \beta_6 & \beta_{10} & \beta_{14} \\ \beta_3 \oplus SNLF[D] & \beta_7 & \beta_{11} & \beta_{15} \end{pmatrix}$$

The 10th round is executed as is, which in turn leads to the following differential characteristics:

$$R_0 \oplus R_1 = \begin{pmatrix} z_0 & 0 & 0 & 0 \\ 0 & 0 & 0 & z_1 \\ 0 & 0 & z_2 & 0 \\ 0 & z_3 & 0 & 0 \end{pmatrix}$$

where $z_0 = S[I_0^{10} \oplus A \oplus SNLF[A]] \oplus S[I_0^{10}]$, $z_1 = S[I_1^{10} \oplus B \oplus SNLF[B]] \oplus S[I_1^{10}]$, $z_2 = S[I_2^{10} \oplus C \oplus SNLF[C]] \oplus S[I_2^{10}]$, and $z_3 = S[I_3^{10} \oplus D \oplus SNLF[D]] \oplus S[I_3^{10}]$.

This is subsequently followed by another infection of R_0 in step 11, captured by the following equations:

$$R_0 \oplus R_1 = \begin{pmatrix} m_0 & 0 & 0 & 0 \\ 0 & 0 & 0 & m_1 \\ 0 & 0 & m_2 & 0 \\ 0 & m_3 & 0 & 0 \end{pmatrix}$$

where $m_j = z_j \oplus SNLF[z_j]$, $j \in \{0, 1, 2, 3\}$. Analogous to the previous round, the current round also has an infection step for R_2, as described next:

$$\begin{pmatrix} \beta_0 \oplus SNLF[A] \oplus SNLF[z_0] & \beta_4 & \beta_8 & \beta_{12} \\ \beta_1 \oplus SNLF[B] & \beta_5 & \beta_9 & \beta_{13} \oplus SNLF[z_1] \\ \beta_2 \oplus SNLF[C] & \beta_6 & \beta_{10} \oplus SNLF[z_2] & \beta_{14} \\ \beta_3 \oplus SNLF[D] & \beta_7 \oplus SNLF[z_3] & \beta_{11} & \beta_{15} \end{pmatrix}$$

In summary, prior to the compulsory dummy round, the number of infected bytes for R_0 and R_2 are found to be 4 and 7, respectively. The final infection occurs in the compulsory dummy round, where the infection spreads across all 16 bytes of R_2, as explained below:

$$T \oplus T^* = \begin{pmatrix} m_0 \oplus 2F_1 \oplus 1F_2 & 1F_3 & 3F_4 \oplus 1F_5 \oplus 1F_6 & 3F_7 \\ 1F_1 \oplus 3F_2 & 1F_3 & 2F_4 \oplus 3F_5 \oplus 1F_6 & m_1 \oplus 2F_7 \\ 1F_1 \oplus 2F_2 & 3F_3 & m_2 \oplus 1F_4 \oplus 2F_5 \oplus 3F_6 & 1F_7 \\ 3F_1 \oplus 1F_2 & m_3 \oplus 2F_3 & 1F_4 \oplus 1F_5 \oplus 2F_6 & 1F_7 \end{pmatrix}$$

where $F_1 = S[\beta_0 \oplus SNLF[A] \oplus SNLF[z_0]] \oplus S[\beta_0]$, $F_2 = S[\beta_{10} \oplus SNLF[z_2]] \oplus S[\beta_{10}]$, $F_3 = S[\beta_3 \oplus SNLF[D]] \oplus S[\beta_3]$, $F_4 = S[\beta_{13} \oplus SNLF[z_1]] \oplus S[\beta_{13}]$, $F_5 = S[\beta_2 \oplus SNLF[C]] \oplus S[\beta_2]$, $F_6 = S[\beta_7 \oplus SNLF[z_3]] \oplus S[\beta_7]$, and $F_7 = S[\beta_1 \oplus SNLF[B]] \oplus S[\beta_1]$.

It is now straightforward to infer the following from the various columns of the state matrix: F_1 and F_2 (column 1), F_3 (column 2), F_4, F_5 and F_6 (column 3), and F_7 (column 4). This allows removal of the effect of the infection caused by the last dummy round. The removal yields the following differential relation:

$$T \oplus T^* = \begin{pmatrix} m_0 & 0 & 0 & 0 \\ 0 & 0 & 0 & m_1 \\ 0 & 0 & m_2 & 0 \\ 0 & m_3 & 0 & 0 \end{pmatrix}$$

The final step is to remove the infection caused by the 10th cipher round. Note that for each $j \in \{0, 1, 2, 3\}$, one can write $z_j \oplus SNLF[z_j]$, where $SNLF[z_j]$ is the result of the infection caused by the Step 11 of Algorithm 10 after the execution of 10th cipher round, while z_j represents the prior differential between the correct and faulty computations. Implementing $SNLF$ as an inversion operation in $GF(2^8)$ yields two potential solutions to z_j for each m_j. Since each equation containing m_j for $j \in \{0, 1, 2, 3\}$ are independent, there are 2^4 possible solutions for $T \oplus T^*$, and hence, T^* as $Tisknown$. Now, once the infection caused by the 10th cipher round has been removed, the adversary simply hypothesizes on 4 bytes of the 10th round key k^{11} and uses the faulty and correct output of 9th cipher round to verify the following relations:

$$
\begin{aligned}
2 \cdot f' \oplus SNLF[2 \cdot f'] &= S^{-1}[T_0 \oplus k_0^{11}] \oplus S^{-1}[T_0^* \oplus k_0^{11}] \\
1 \cdot f' \oplus SNLF[1 \cdot f'] &= S^{-1}[T_{13} \oplus k_{13}^{11}] \oplus S^{-1}[T_{13}^* \oplus k_{13}^{11}] \\
1 \cdot f' \oplus SNLF[1 \cdot f'] &= S^{-1}[T_{10} \oplus k_{10}^{11}] \oplus S^{-1}[T_{10}^* \oplus k_{10}^{11}] \\
3 \cdot f' \oplus SNLF[3 \cdot f'] &= S^{-1}[T_7 \oplus k_7^{11}] \oplus S^{-1}[T_7^* \oplus k_7^{11}]
\end{aligned}
$$

where $SNLF[b \cdot f']$ for $b \in \{1, 2, 3\}$ represents the outcome of the infection caused in Step 11, after the 9th cipher round. These equations are solved to obtain all 2^4 possible values of T^*. Note that every solution of T^* gives 1036 candidate values for 4 bytes of the 10th round key k^{11} as described in [147]. Thus the expected number of candidate values for 4 bytes of k^{11} is $2^4 * 1036 = 16576$. Repeating the attack on a second pair of correct and faulty ciphertexts brings down the number of possible values for 4 bytes of k^{11} to only two [147].

10.4.4 Attack in the Presence of Multiple Dummy Rounds

If the adversary were to launch the aforementioned attack in the presence of random intermediate dummy rounds, the most profitable strategy is to inject the fault towards the end of the encryption operation. For example, suppose that the total number of rounds for which a single encryption runs is $22 + d$, where d is the number of dummy rounds. Then, the execution flow of Algorithm 10 is symbolically determined by a binary string of length $22 + d$, with exactly 22 entries that are 1, and the remaining entries being 0. Now, consider the probability that the $(22 + d - 2)$th *RoundFunction* is a 9th cipher round. This is the same as the probability that the binary string ends with three consecutive 1's. The probability of this event is given by $\frac{(19+d)!/((19)!\cdot(d)!)}{(21+d)!/((21)!\cdot(d)!)}$. If $d = 20$ then the probability that 40th *RoundFunction* is a 9th cipher round is nearly 0.26. Hence, the adversary indeed only needs to execute a greater number of fault injection trials to achieve the desired faulty ciphertexts for the attack.

10.5 The Second Infective Countermeasure

In this section, we present a second infective countermeasure, introduced by Tupsamudre et al. in [173], that overcomes the loopholes of the infective countermeasure of Gierlichs et al. [70]. As evident from the attacks explained earlier, the infective countermeasure for protecting AES against the differential fault attacks should have the following properties:

1. If a fault is injected in any of the cipher, redundant or dummy round, all bytes in the resulting ciphertext should be infected.

2. The infection technique should nullify any attempts to make hypothesis on the secret key used in AES from the faulty ciphertext.
3. All *RoundFunction* computations (cipher, redundant or dummy rounds) should be indistinguishable from a side-channel perspective.

The infective countermeasure proposed by Tupsamudre et al. is designed to meet these guidelines. The basic principle of interspersing cipher and redundant rounds with dummy computational rounds remains unaltered. However, the countermeasure is more robust in its response to fault attacks, as opposed to the previous countermeasure. Algorithm 11 summarizes this countermeasure scheme.

Algorithm 11 The Second Infective Countermeasure

Input: P, k^j for $j \in \{1, \ldots, n\}$, (β, k^0), $(n = 11)$ for AES-128

Output: C = BlockCipher(P, K)

State $R_0 \leftarrow P$, Redundant state $R_1 \leftarrow P$, Dummy state $R_2 \leftarrow \beta$

$i \leftarrow 1, q \leftarrow 1$

$rstr \leftarrow \{0, 1\}^t$ // $\#1(rstr) = 2n, \#0(rstr) = t - 2n$

while do$q \leq t$ do:

  $\lambda \leftarrow rstr[q]$ // $\lambda = 0$ implies a dummy round

  $\kappa \leftarrow (i \wedge \lambda) \oplus 2(\neg\lambda)$

  $\zeta \leftarrow \lambda \cdot \lceil i/2 \rceil$ // ζ is actual round counter, 0 for dummy

  $R_\kappa \leftarrow RoundFunction(R_\kappa, k^\zeta)$

  $\gamma \leftarrow \lambda(\neg(i \wedge 1)) \cdot BLFN(R_0 \oplus R_1)$ // check if i is even

  $\delta \leftarrow (\neg\lambda) \cdot BLFN(R_2 \oplus \beta)$

  $R_0 \leftarrow (\neg(\gamma \vee \delta) \cdot R_0) \oplus ((\gamma \vee \delta) \cdot R_2)$

  $i \leftarrow i + \lambda$

  $q \leftarrow q + 1$

end while

return R_0

Following additional notations are used in this algorithm:

1. **rstr**: A 't' bit random binary string, consisting of $(2n)$ 1's corresponding to AES rounds and $(t - 2n)$ 0's corresponding to dummy rounds.
2. **BLFN**: A boolean function that maps a 128 bit value to a 1 bit value. Specifically, $BLFN(0) = 0$ and for nonzero input $BLFN$ evaluates to 1.
3. $\boldsymbol{\gamma}$: A one bit comparison variable to detect fault injection in AES round.
4. $\boldsymbol{\delta}$: A one bit comparison variable to identify a fault injection in dummy round.

The main features of Algorithm 11 that makes it stronger than Algorithm 10 are:

1. Multiple dummy rounds can occur even after the execution of last cipher round. Hence, the 10th cipher round is not always the penultimate round.
2. The state matrix R_2 for the dummy round is a uniformly random value β that is completely independent of the intermediate cipher state for either the redundant or the cipher rounds. Hence, in the event of an infection, the faulty ciphertext is completely random, which makes DFA impossible.

Table 10.1 Computation of Algorithm 11

Step	Redundant round	Cipher round	Dummy round
6	$\lambda = 1$, i is odd	$\lambda = 1$, i is even	$\lambda = 0$
7	$\kappa \leftarrow 1$	$\kappa \leftarrow 0$	$\kappa \leftarrow 2$
8	$\zeta \leftarrow \lceil i/2 \rceil$	$\zeta \leftarrow \lceil i/2 \rceil$	$\zeta \leftarrow 0$
9	$R_1 \leftarrow RoundFunction(R_1, k^{\zeta})$	$R_0 \leftarrow RoundFunction(R_0, k^{\zeta})$	$R_2 \leftarrow RoundFunction(R_2, k^0)$
10	$\gamma \leftarrow 0$	$\gamma \leftarrow BLFN(R_0 \oplus R_1)$	$\gamma \leftarrow 0$
11	$\delta \leftarrow 0$	$\delta \leftarrow 0$	$\delta \leftarrow BLFN(R_2 \oplus \beta)$
12	$R_0 \leftarrow R_0$	$R_0 \leftarrow (\neg(\gamma) \cdot R_0) \oplus ((\gamma) \cdot R_2)$	$R_0 \leftarrow (\neg(\delta) \cdot R_0) \oplus ((\delta) \cdot R_2)$
13	$i \leftarrow i + 1$	$i \leftarrow i + 1$	$i \leftarrow i + 0$
14	$q \leftarrow q + 1$	$q \leftarrow q + 1$	$q \leftarrow q + 1$

3. Leakage via timing side-channels is prevented by keeping the number of dummy rounds constant.

For a clear illustration, Table 10.1 shows the functioning of Algorithm 11.

We leave it an an exercise to the reader to formally verify the correctness of the algorithm under undisturbed execution.

10.6 Summary

In this chapter, we presented to the readers two state-of-the-art *infective countermeasure* techniques to protect AES-128 against DFA. We first presented a countermeasure from Gierlichs et al., and showed some of its vulnerabilities against DFA attacks. We then presented a second countermeasure by Tupsamudre et al., that overcomes these vulnerabilities.

Chapter 11
Reactive Design Strategies Against Fault Injection Attacks

Jakub Breier, Wei He and Shivam Bhasin

Abstract In this chapter, we present to the readers two fault injection sensors against powerful fault injection techniques like laser and electromagnetic glitch. The first sensor is built upon a watchdog ring oscillator monitor with a standard Phase Locked Loop (PLL) block. Owing to the scarcity of PLL in resource-deficient devices, an all-digital alternative is proposed. The sensors are designed to be more reactive to fault injection methods than the sensitive target, in order to raise an alarm with solid security margin. Both countermeasures are low-cost and can be used to protect any arbitrary target circuit. Practical validation against near-infrared laser on Xilinx Virtex-5 FPGA is provided.

11.1 Introduction

A typical countermeasure against the fault injection attacks is to modify the algorithm, to make it possible to detect the logic abnormality. These methods can be redundant multiple rails [164] and parity based error detection [84]. However, these solutions do not provide any security margin to foresee an upcoming injection in terms of time or energy. Besides, the architectural modification of the original circuit to be protected is not applicable to the hardened IP cores, and also inevitably degrade the performance of the original algorithm. The preferable method for countering the injection attack is to integrate a sensor logic which should be more sensitive to the electrical impact from the injection than the circuit itself, in order to detect the threat earlier. In this way, a defense system can react to the injection before a value fault has really been injected into the protected circuit. Moreover, the countermeasure

J. Breier (✉) · S. Bhasin
Physical Analysis and Cryptographic Engineering, Temasek Laboratories, Nanyang Technological University, Singapore, Singapore
e-mail: JBreier@ntu.edu.sg

W. He
Shield Lab, Central Research Institute, Huawei International Pte. Ltd., Singapore, Singapore

S. Patranabis and D. Mukhopadhyay (eds.), *Fault Tolerant Architectures for Cryptography and Hardware Security*, Computer Architecture and Design Methodologies, https://doi.org/10.1007/978-981-10-1387-4_11

should be standalone, hence it can be implemented independently from the protected algorithm, without some major modifications to the algorithm core itself.

In this chapter, we examine error detection possibilities based on hardware-level sensors. Both analog and digital sensors are presented and evaluated in a real-world setting, by using an FPGA implementation of PRESENT block cipher. The evaluation was done by using one of the most powerful and precise fault injection techniques – laser fault injection.

The rest of this chapter is structured as follows. Section 11.2 provides a necessary technical background for this work. Sections 11.3 and 11.4 present analog and digital countermeasures for detecting faults, respectively. Experimental evaluation of these approaches is stated in Sect. 11.5, and finally, Sect. 11.6 summarizes this chapter.

11.2 Technical Background

This section provides basic background on architecture of modern FPGA (target platform). Please note that the readers are expected to be familiar with the structure and specifics of the lightweight block cipher PRESENT, which has already been described in Chap. 4.

11.2.1 FPGA Architecture

Modern FPGA architecture consists in a vast logic cell pool where each fundamental cell is deployed inside a regular logic array. To implement a designed circuit, the vendor tools will synthesize the native high-level described design into a netlist which only contains the very fundamental logic gates and this netlist is further processed to be compatible with the specific device according to the feature for logic cell of each specific device. For instance, the gates will be complied and re-merged to form individual look-up-tables (LUTs) that each can be mapped to the available LUT resources in cells. After this step, the circuit netlist is routed to connect the circuit, populating as a complete network. This place and route steps will be repetitively done if the timing, area, power or other user-imposed constraints are not satisfied. Eventually the bitstream can be created to be configured into the target programmable device.

One of the primary FPGA vendors today is Xilinx, well-known for their wide range of products for both academic and industrial applications. While it is true that modern-day technology advances rapidly, which requires regular product upgradation and enhancement of product functionality, the fundamental underlying architecture is usually agnostic of such changes, and remains the same across most devices. One such architecture for a Xilinx Virtex-5 FPGA is illustrated in Fig. 11.1. The basic architecture comprises of a mesh-like array of configurable logic blocks (CLBs). A CLB is composed of two slices, each in turn containing four multiplexers, four 6-input

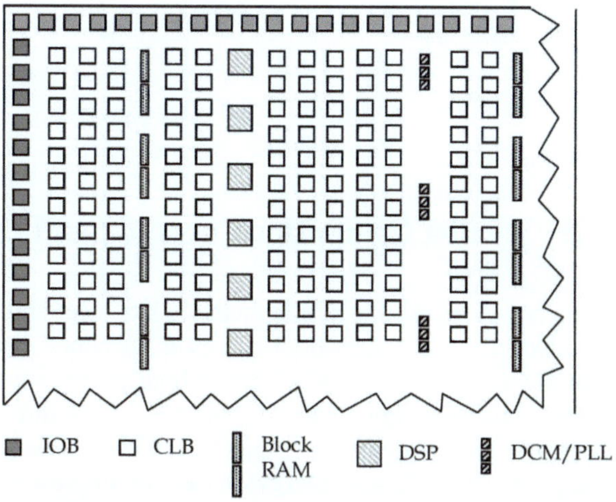

Fig. 11.1 Virtex-5 architectural overview [76]

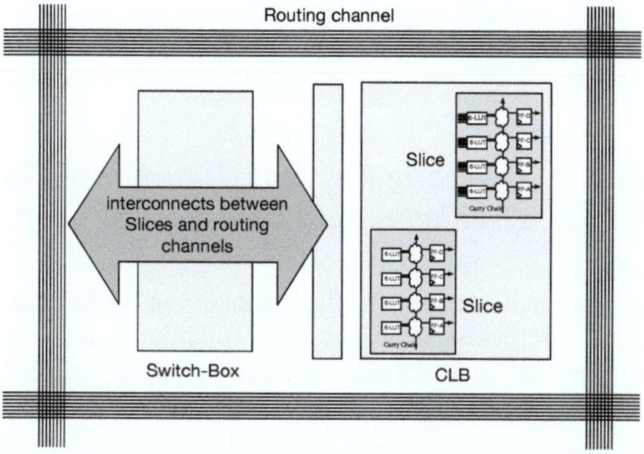

Fig. 11.2 CLB logic cluster and routing resources [76]

look-up-tables (LUTs), four flip-flops a single carry chain. An LUT is essentially a 64-bit SRAM that can be used to implement a variety of Boolean functions under appropriate configuration. These include a solitary 6-input Boolean function, two Boolean functions with no more than five shared inputs, and two Boolean functions with no more than five different inputs. Additionally, each FPGA is also equipped with additional on-chip functional modules, including (but not limited to) a phase locked loop (PLL), a digital clock manager (DCM), system monitors for temperature and voltage, multiple memory elements such as Block RAMs and digital signal processor (DSP) blocks (Fig. 11.2).

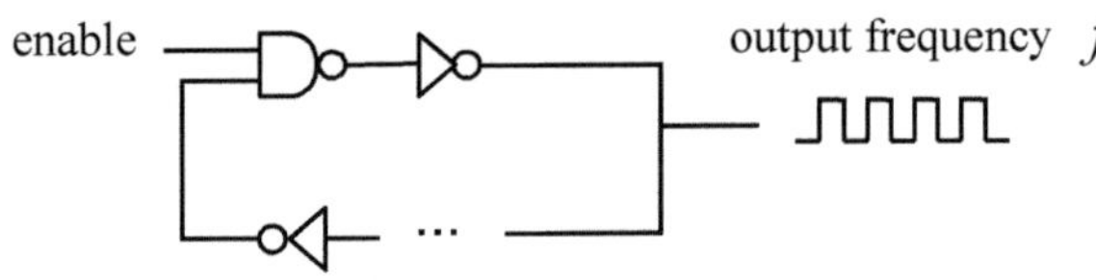

Fig. 11.3 Digital inverter based ring-oscillator

11.3 Analog Countermeasures to Fault Detection

When advanced fault injection methods like laser and EM are used, the intention of the adversary is to inject high energy disturbance into the device to induce faults. In principle, detecting any deliberate high energy injection can inform on attempts of fault injection. For instance, photodetectors and equivalent custom cells were previously used to detect laser injection. If the a fault injection device is well focused to a particular cell, this energy can cause single event upsets. This part explores the potential of an RO to detect energy induced in the integrated circuit by using readily available components and cell library to design a countermeasure, specially for FPGA.

11.3.1 Detecting Physical Disturbance Through Ring Oscillator and Phase-Locked Loop

A ring oscillator (RO), as illustrated in Fig. 11.3, has a closed loop construction with a chain of inverters, subject to the condition that the number of inverters is odd. Any RO operates in two phases - a warm-up phase just after start-up, to lock upon a stable state frequency f_{RO} (which is essentially the inverse of the RO path delay t_{RO}), and a functional phase where the oscillator outputs are generated. The state frequency f_{RO} depends on a number of factors, most notably process, voltage and temperature (PVT) variation.

The main idea of using an RO to detect laser fault injections is as follows: the energy induced by a laser pulse instantaneously modifies the PVT parameters of an RO located near the impact region; consequently, it displays easily discernible alterations in the frequency and phase of the output clock signal. The phenomenon is experimentally illustrated in Fig. 11.4, where a clock signal is generated by a single RO, implemented across 9 adjacent CLBs of an FPGA. The default frequency of the clock signal so generated in 210 MHz (see Fig. 11.4 (top)), while laser injection results in an instant modification of the PVT parameters, leading to alterations in the frequency, amplitude and phase of the clock signal ((see Fig. 11.4 (bottom))). On the other hand, removing the laser pulse causes the clock signal to retract to its normal operating frequency, implying that the PVT parameters of the RO have returned to normal. The behavior of the RO under the impact of EM pulse injections is found to be similar.

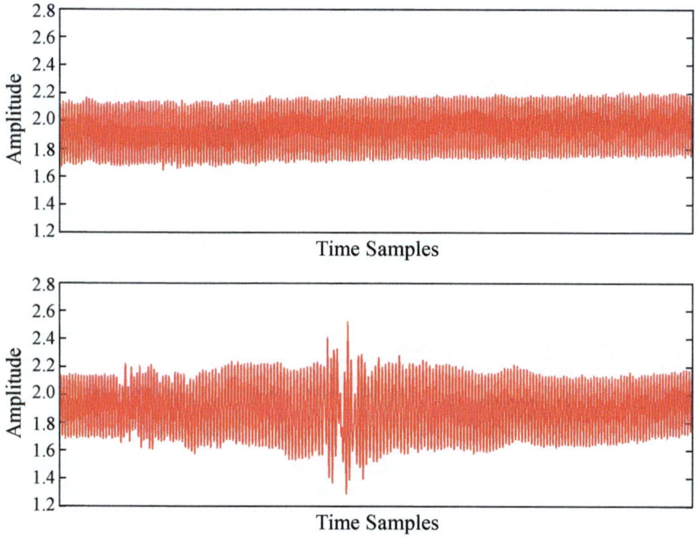

Fig. 11.4 RO generated clock (top) under impact of laser injection (bottom) [76]

The main challenge here is to detect the impact of the laser fault injection on the output clock generated by the RO. For this, we use a phase-locked loop (PLL) block, typically available on any modern FPGA. A PLL is essentially a clock control circuit, which phase synchronizes a clock source and a feedback signal via monitoring their frequency and phase differences. It has three main components - a Phase-Frequency Detector (PFD), a Low Pass Filter (LF), and a Voltage-Controlled Oscillator (VCO). As demonstrated in Fig. 11.5, the polarity and magnitude of the differences between the input and feedback clocks are measured by the PFD unit, and converted into *UP* and *DOWN* pulses. The LF unit then converts these pulses into a control voltage signal to drive the VCO and achieve phase and frequency synchronization.

The overall countermeasure strategy is thus fairly simple: the clock signal generated by an RO module acts as our indicator. We continuously monitor the this signal using a PLL block. Under normal operating conditions, the RO would produce a clock signal with a stable output frequency, amplitude and phase (neglecting local jitters), and the PLL would be in a locked state (Fig. 11.5a). Any sudden alterations in the frequency, phase or amplitude of the clock signal would force the PLL into an unlocked state (Fig. 11.5b), and an alarm is raised to indicate the presence of deliberate energy injections due to laser/EM pulses. This alarm signal can then be used to trigger countermeasure schemes such as core shutdown, enabling additional randomizations etc. The RO module is placed directly above the sensitive crypto core for easy detection. Most importantly, the countermeasure strategy is extremely lightweight, with an area overhead of **a single LUT for the RO module, a single PLL block, a single DFF for the alarm** and any additional routing resources.

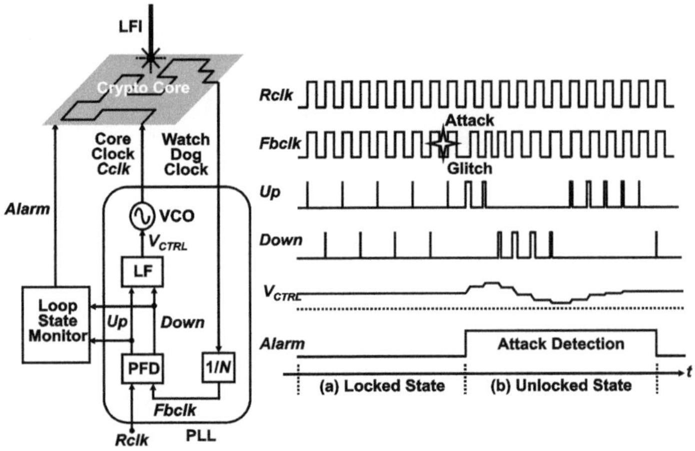

Fig. 11.5 Block diagram of PLL-based FI countermeasure. The waveforms depicts **a** lock and **b** unlock state [76]

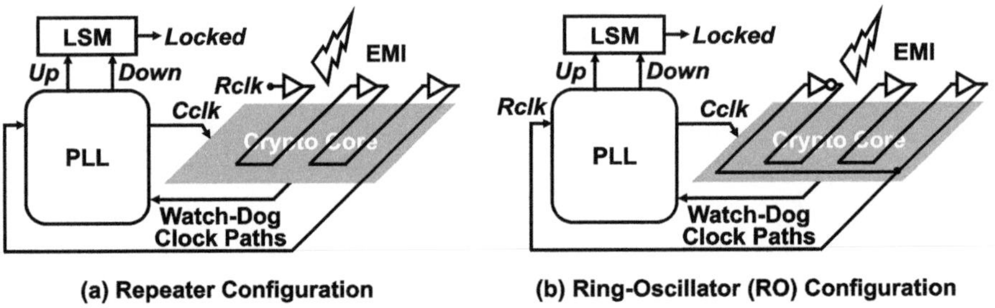

Fig. 11.6 Security-enhanced implementation of EMI countermeasure with reference clock watch dog in **a** repeater and **b** ring-oscillator configuration [122]

The aforementioned detection strategy can be further enhanced by amalgamating the reference clock paths with the already existing watch-dog clock paths (Fig. 11.6). This is a pre-detection strategy, implying that it can detect fault attacks even prior to the actual fault injection in the target crypto core. In addition, it can also detect fault injections in the control circuit of the PLL. Figure 11.6a illustrates a repeater configuration of the countermeasure strategy, wherein synchronization with the external clock source *Rclk* is flexible, but the clock port is exposed to the adversary. In reality, attacking clock ports is hard, since attempting to insert clock glitches also causes the PLL to reach an unlocked state and raises a fault detection alarm. The strategy can be slightly tweaked to also prevent enhanced fault analysis techniques such as fault sensitivity analysis (FSA) [112] discussed previously. For example, one could set the critical timing delay of the PLL block to be lower than that for the crypto core; in this case, the first alarm is raised for the faulty operation of the PLL core, and the critical fault injection intensity for the actual crypto core is masked from the adversary.

Finally, Fig. 11.6b represents the alternative RO-based configuration which additionally hides the clock port and maximizes the security guarantees. All communication between the crypto core and other modules happens via asynchronous hand-shaking protocols.

11.3.2 Implementation Constraints of Ring Oscillator

As RO is the core of presented fault detection sensor, the implementation aspects of RO should be carefully investigated. A basic RO is illustrated in Fig. 11.3 which is composed of odd number of inverters. Modern CAD tools are designed to find any redundancy in a given design and optimize it for area and performance. Such tools can easily detect that odd number of inverters and optimize it to keep a single inverter. This optimization would result in functional equivalence but other timing and frequency might not be required. For instance, by increasing and decreasing the number of inverters, the oscillation frequency of the RO can be controlled. Thus proper design constraints must be put in place to prevent unwanted optimization of the RO. An automated method to design desired RO in Xilinx FPGA was shown in [122].

Although the proposed fault detection sensor using PLL will work against different fault injection methods, the implementation of RO must be adjusted to the kind of attack. The energy injection in laser stays very localized, which makes it suitable for SEU. Therefore, the RO must be placed close to point of impact, otherwise the energy might not diffuse enough to be detected by RO sensor. If the placement of RO is not precise, the RO might have low or even zero detection rate. This is specially true for high laser power when the beam is focused and suffer little or no diffraction. If the sensitive circuit to protect is large, it is better to deploy several small sensors to achieve a higher detection rate. The RO restriction for EM fault injection are different. As disturbances are created using EM field in case of EM fault injection, the impact area of EM field are generally bigger than focused laser spots. Thus, a single sensor with bigger RO encapsulating the sensitive circuit can already provide a good detection rate. Another trick to increase the detection rate of the sensor is to route the RO in a arbitrary manner. By choosing arbitrary routing over the sensitive circuit, rather than normal rectangular, the attack detection surface is increased. For instance in [122], authors proposed a zig-zag routing to fully encapsulate the sensitive circuit. The number of turns or nodes in the routing is determined by the desired oscillating frequency.

11.4 Digital Countermeasures to Fault Detection

A fault injection sensor using PLL was proposed in the previous section. The proposed sensor is low-cost as composed of a simple RO (watchdog) and a (monitoring) PLL block. As the sensor stays independent of underlying circuit to protect, the sensor is easy to integrate with different designs. The only aspect which needs careful designer intervention is the constraints on RO. Despite all these advantages, a major limitation of the proposed sensor is the use of PLL block. PLL is a commonly used clock control circuit but the availability of PLL is not always guaranteed. Most modern FPGA are equipped with a PLL blocks but the number of PLL blocks are scarce and might already be completely used by the main application, leaving no scope for integration of the proposed sensor. Older FPGA might not support PLL. The use of PLL also reduces the portability of the proposed sensor from FPGA to ASIC, where PLL is not generally part of the standard library and either must be custom designed or purchased as an IP. Availability of PLL is a also a strong assumption for low-cost applications like internet of things (IoT). To overcome these limitation, we propose an all-digital solution to design the fault injection sensor.

11.4.1 Digital Fault Injection Detector

In this section, we present to the reader a digital fault injection detector. An overview of the detection mechanism is provided in Fig. 11.7. The main component of the scheme is a multi-inverter ring oscillator (RO) serving as the `Watchdog Sensor` for sudden alterations to the RO frequency. The other component is the `Disturbance Capture` logic comprised of two flip-flops and a logic gate i.e., ($Q1\&\overline{Q2}$) [75]. Two flip-flops FF1 and FF2 are used to sample the frequencies from two different points - *f1* and `f2` on the RO loop. The working of the detection system is summarized in Fig. 11.8. Quite evidently, the RO module comprises of three inverters, the outputs of which are named as *f1*, *ck*, *f2* in the sequence of signal propagation. The negative edge of *ck* triggers the flip-flops FF1 and FF2 in the `Disturbance Capture` module, as shown in Fig. 11.8a. Under stable operating conditions (that

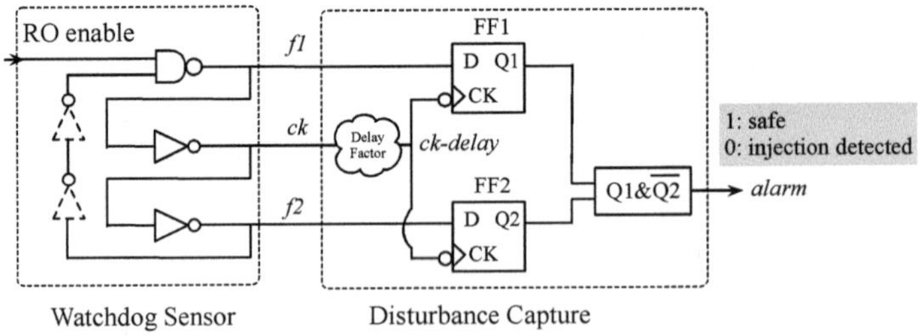

Fig. 11.7 Topology of the schemed fault injection sensor system

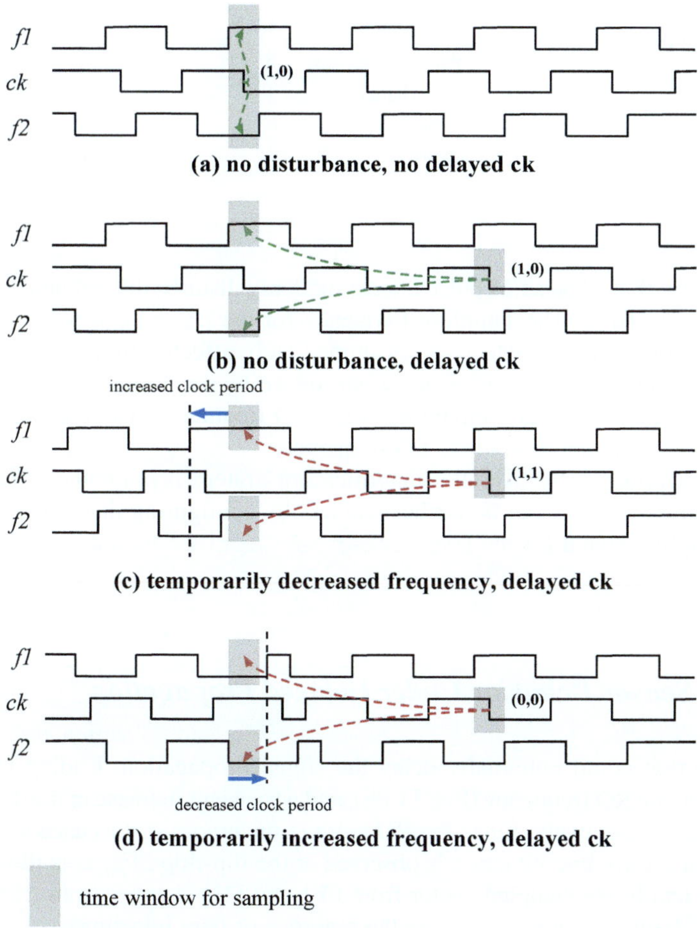

Fig. 11.8 Timing of low/high-frequency ripple detection

is, in the absence of external pulses), the three signals have the same phase shift, while the polarity of *ck* is opposite to that of *f1* and *f2*. Additionally, when there is no signal delay from the RO to FF1 and FF2, their sampled values are `'1'` and `'0'`, respectively (see the blue dotted arrow lines in Fig. 11.8a). Any ripples in the RO output affects the frequencies of all three inverters, and are hence not captured by the `Disturbance Capture` module.

Now, how do we capture the disturbances to the RO inverted frequencies introduced by fault injection? The trick is to insert an additional `delay factor` into the clock inputs of the flip-flops FF1 and FF2. This in turn implies that instead of being triggered by the falling edge of the original *ck* clock signal, the flip-flops are now triggered by the falling edge of a different signal *ck-delay*, which is delayed with respect to *ck* by several clock cycles (see the red dotted arrow lines in Fig. 11.8a).

Table 11.1 Area report of the all-digital fault injection sensor

Component	LUT	DFF
Watchdog sensor	3	0
Disturbance capture	1	2
Delay	1	0

Now suppose that a laser/EM pulse injects suddenly disturbs the frequency/phase of the signals *f1* and *f2*. The sampling frequency for the `Disturbance Capture` module is now *ck-delay*, which is not immediately affected by the fault injection due to the additional propagation delay introduced by the designer. This allows detection of bi-directional abnormalities in the RO frequency, which could be either acceleration or deceleration of the RO frequencies.

The area report for the overall countermeasure strategy is presented in Table 11.1. Note that the overhead can be further reduced by configuring the additional delay in the sampling signal for the `Disturbance Capture` module via appropriate routing only.

11.4.2 Sensor Behavior Under Delayed Propagation

Fault injection could potentially delay the signal propagation, leading to a short reduction in the RO frequency (Fig. 11.8b), and a transient increase in the duty cycles of *f1* and *f2*, respectively. Since the RO module undergoes disturbance due to fault injection, a set-up time violation is observed at the flip-flop FF2, as is illustrated in Fig. 11.8a and b (the sampled vector from FF1 and FF2 changes from `'10'` in the absence of fault injection to `'11'` in the presence of fault injection).

On the other hand, fault injection could also lead to a transient increase in the operating frequency of the RO module. In this reverse scenario, the duty cycles of *f1* and *f2* decrease instead of increasing, leading to set-up time violations at FF2 instead of FF1. See, for example, Fig. 11.8c, where the sampled value vector from FF1 and FF2 changes from `'10'` to `'00'` instead of `'11'`.

11.4.2.1 Sensor Behavior Under Complex Disturbances

In real life scenarios, the effect of fault injection is rarely limited to simple transient acceleration or deceleration of the RO propagation delay within a single clock cycle. The effect of the injection could potentially span several clock cycles, and comprise of random acceleration and delay phases. However, note that in the aforementioned countermeasure strategy, the first glitch in the disturbance period raises an alarm. Also, a complex disturbance can be viewed as a collection of simple single-cycle disturbances, the first of which is guaranteed to be captured by the countermeasure strategy. In other words, complex disturbances arising due to fault injection do not impede detection via simple countermeasure strategies.

11.5 Experimental Results

In this section, we present to the reader a prototype implementation of the sensor-based countermeasure, and practically evaluate the security of the same against laser fault injection attacks. The implementation platform is a Virtex-5 FPGA, and the target implementation is a PRESENT block cipher, implemented in iterated fashion with a 64 bit data register. On this platform, a CLB array is implemented. As already mentioned, a Virtex-5 FPGA is already equipped with the other necessary modules such as PLLs, DSP blocks and DCM modules. Interconnections between the CLB logic and the external routing channels are provisioned via additional switch-boxes. Note that on a Virtex-5 FPGA, a single CLB comprises of two slices, with each slice composed of four LUTs (that implement all synthesized logic gates), and four flip-flops. In particular, the 64-bit state register of PRESENT is implemented using these flip-flops.

11.5.1 Experimental Setup

The overall implementation of the PLL-based detection strategy is depicted in Fig. 11.9 on a Xilinx Virtex-5 (VLX50T) FPGA, manufactured by 65 nm technology with a `flip-chip` package. The 64-bit `state registers` of the PRESENT-80

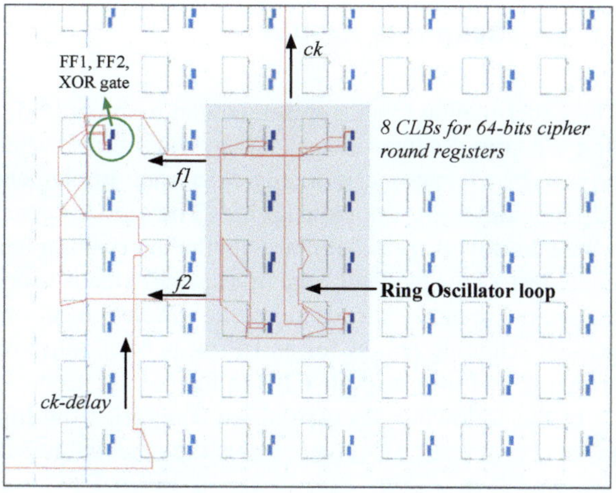

Fig. 11.9 FPGA implementation scheme of the proposed sensor system and the protected 64-bit round data registers of the PRESENT-80 cipher

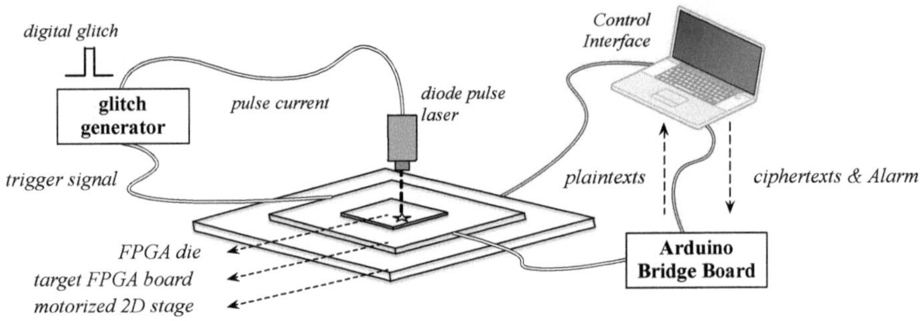

Fig. 11.10 Illustration of LFI experimental setup

cipher are implemented across a rectangle of 16 slices (4 flip-flops per slice), which requires a total of 8 CLBs. Additionally, all the protected data registers are encompassed by forcing the RO routing path to cross all the four corners of this rectangle. On the other hand, the `Disturbance Capture` module, which takes as input the RO inverter signals, is deployed outside this path. In order to allow for partial absorption of energy from the laser photons by the chip substrate, the substrate is milled down to roughly 130 μm. This allows the laser photons to reach the target logic without losing too much energy. The FPGA board (Digilent *Genesys*) is fixed on a motorized 2-dimensional (X-Y) stepper stage, with 0.05 μm minimum step size. The FPGA logic core is connected to a controller GUI on a PC via an `Arduino Due` board. This allows us to observe and record the encryption outputs and the alarm signal in real time, the `Arduino Due` board is programmed to connect the overall setup is depicted in Fig. 11.10. The injection was made very precise by reducing the spot size from a diode pulse laser with 1064 nm wavelength, to 60×14 μm via a $5\times$ magnification.

11.5.2 Scanning Results

The experimental results presented can be broadly divided into two scenarios. The first scenario assumes a weak adversarial model, where the target device is a black box to the adversary and details of internal implementation and architecture of the target modules are unknown apriori. The second scenario assumes a stronger model where the adversary is familiar with the internal implementation details of the device under test. Quite evidently, the first scenario leads to coarse-grained injections, while the second scenario affords more fine-grained injection capabilities. In particular, we present scanning results for an LFI scan over entire regional CLB array in the first scenario, and over a single CLB in the second scenario.

In the following discussion, we categorize the injected faults into three types: (a) `Only Alarm` (F_{OA}) represents the detected injection without cipher faults; (b)

`Fault + Alarm` (F_{FA}) refers to the detected injections that induced cipher faults, and (c) `Only Fault` (F_{OF}) denotes the injections that induced cipher faults without triggering the alarm. We denote by `Detection Rate` of a sensor the ratio $\frac{F_{FA}}{F_{FA}+F_{OF}}$ and by `Alarm Rate` the ratio $\frac{F_{OA}+F_{FA}}{F_{FA}+F_{OF}}$. The former is an indicator of the fraction of fault injections into the main cipher logic that are detected, while the latter indicates is an indicator for the fraction of all fault injections that raise an alarm upon injection in the system.

11.5.2.1 Regional Scan

The first scenario that we present to the readers assumes that the device under test is a black-box to the adversary. In order to identify the sensitive locations for potential fault injection, the adversary would need to perform a coarse surface scan over a big fabric region, that would typically encompass not only the cipher data registers circumvented by the RO routing path, but also some other neighboring regions of the device. In our experiments, the scan matrix was therefore chosen to have dimensions 300 × 300, with a single injection per point (a total of 90,000 injections). Figure 11.11 compares the outcome of LFI scans on the two countermeasure implementations - the first using the analog PLL-based sensor and the other using the digital RO-based sensor. Quite evidently, from Fig. 11.11a, the PLL-based sensor managed to detect injections not only in the RO region but also on other CLBs in the vicinity ($F_{OA} = 271$). It has a `Detection Rate` of 100% and an `Alarm Rate` of 91.33:1. Figure 11.11b, on the other hand, demonstrates that the RO-based digital sensor has a much higher `Alarm Rate` of 678.63:1:1, while retaining a 100% `Detection Rate`. This readily shows that the digital sensor has a much higher alarm density, and hence much greater sensitivity to fault injection attacks, as compared to its analog counterpart. Also, for the digital sensor, the lowest power

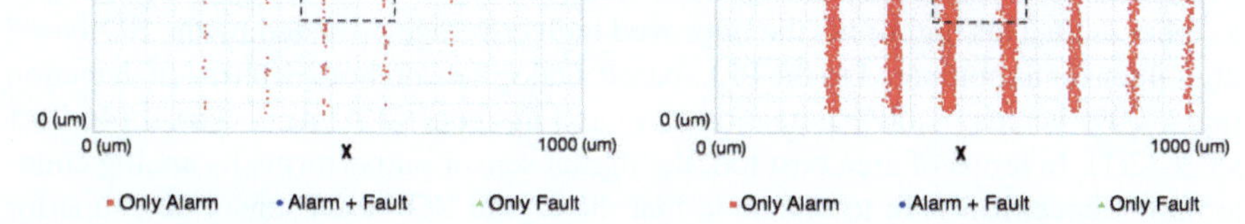

Fig. 11.11 Laser fault injection scan to regional silicon **a** PLL based LFI sensor; **b** the proposed digital LFI sensor

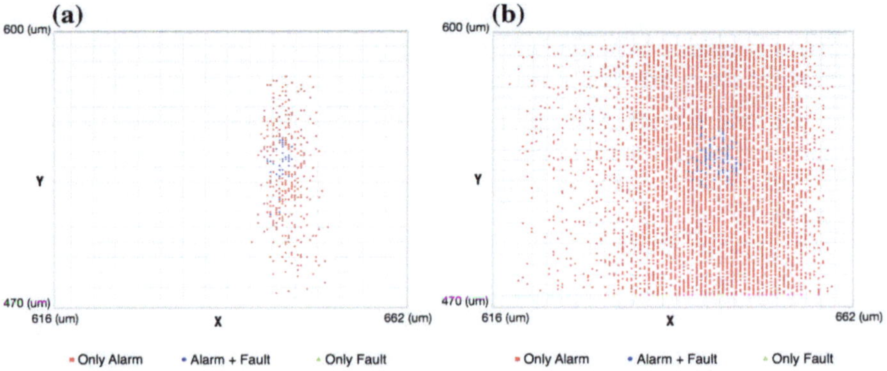

Fig. 11.12 Laser fault injection scan to a single CLB: **a** PLL based LFI sensor; **b** the proposed digital LFI sensor

level to trigger an alarm is 44% of the full laser strength, while the lowest power level to precisely inject a fault in the cipher state is 75% of the full laser strength. This allows a security margin of 31%. Refer the upper part od Table 11.2 for the detailed comparison results.

11.5.2.2 Single-CLB Scan

We present to the reader a second scenario that assumes greater adversarial capabilities. The adversary is assumed to be familiar with the internal implementation details and architecture of the target device, and can therefore precisely target the laser beam on the CLB inside the chip, where the sensitive internal state registers of the PRESENT-80 are located. Recall that each CLB has 4 registers implemented inside; the adversary is assumed to target one of these CLBs. Since the laser is now highly focused, the chances of fault injection are naturally higher. In particular, the scan matrix used by the adversary can be much smaller (150 × 150), with a single injection per point. The results for fault injection experiments are presented in Table 11.2. The PLL sensor successfully detected 284 injections resulting in no cipher faults and 33 injections resulting in cipher faults, while a single injection leading to a cipher fault went undetected. The digital sensor, on the other hand, detected as many as 4461 injections resulting in no cipher faults and all 99 injections leading to a cipher fault. This highlights the improved fault coverage provided by the RO-based digital sensor as compared to the PLL-based analog sensor, both in terms of detection rate (100% versus 97.06% detection rate) and the general `Alarm Rate` (46.06:1 vs. 9.32:1). In terms of area cost too, the digital sensor outperforms its analog counterpart. Hence, it is safe to conclude that the digital RO-based sensor is a superior alternative to the analog PLL-based sensor against LFI (Fig. 11.12).

Table 11.2 Experimental results comparison between the PLL based sensor and the presented digital sensor using LFIs

		Only alarm F_{OA}	Fault+Alarm F_{FA}	Only fault F_{OF}	Scan matrix	RO freq. (MHz)
PLL LFI sensor (Region scan)	No.	271	3	0	300 × 300	≈220
	Min.Power	54%	90%	*n/a*		
	Detection	Successful		Failed		
	Detection rate	$\frac{F_{FA}}{F_{FA}+F_{OF}} = 100\%$				
	Alarm rate	$\frac{F_{OA}+F_{FA}}{F_{FA}+F_{OF}} = 91.33 : 1$				
Digital LFI sensor (Region scan)	No.	5421	8	0	300 × 300	≈206
	Min.Power	44%	75%	*n/a*		
	Detection	Successful		Failed		
	Detection rate	$\frac{F_{FA}}{F_{FA}+F_{OF}} = 100\%$				
	Alarm rate	$\frac{F_{OA}+F_{FA}}{F_{FA}+F_{OF}} = 678.63 : 1$				
PLL LFI sensor (CLB scan)	No.	284	33	1	150 × 150	≈220
	Min.Power	60%	75%	*n/a*		
	Detection	Successful		Failed		
	Detection rate	$\frac{F_{FA}}{F_{FA}+F_{OF}} = 97.06\%$				
	Alarm rate	$\frac{F_{OA}+F_{FA}}{F_{FA}+F_{OF}} = 9.32 : 1$				
Digital LFI sensor (CLB scan)	No.	4461	99	0	150 × 150	≈206
	Min.Power	42%	63%	*n/a*		
	Detection	Successful		Failed		
	Detection rate	$\frac{F_{FA}}{F_{FA}+F_{OF}} = 100\%$				
	Alarm rate	$\frac{F_{OA}+F_{FA}}{F_{FA}+F_{OF}} = 46.06 : 1$				

11.5.3 Full Cipher Protection

While the watchdog RO covers a total of 24 CLBs, fault injections to its neighboring CLBs also raise the alarm, as depicted in Fig. 11.11; thus the RO effectively covers a sufficiently wide fabric region. To test this, two PRESENT-80 block cipher modules were implemented in parallel, with the aim of covering the logic resources in as large an area as possible (see Fig. 11.13). This makes the implementation more vulnerable to fault attacks.

In this experiment, the LFI scan matrix targeting the dual cipher implementation was expanded to 300×300 points, with a single injection peer point. As demonstrated in Fig. 11.14, the fault distribution shows fewer observed faults (in part due to the lower injection density). In all, 65 out of 69 injected faults raised an alarm, while only 4 fault instances went undetected. This gives a `Detection Rate` of 94.20%. Also, besides the valid 65 instances, there were 116 other instances when the alarm was raised despite the fault injection not being in the cipher module. This gives an `Alarm Rate` of 2.63:1, meaning that even when the cipher implementation is very large and highly dense, the chances of setting off the alarm are 2.63 times higher than successfully injecting a fault in the target location. Certain outlier faults that were observed on the I/O and power pad; but these do not threaten the target cipher in any way.

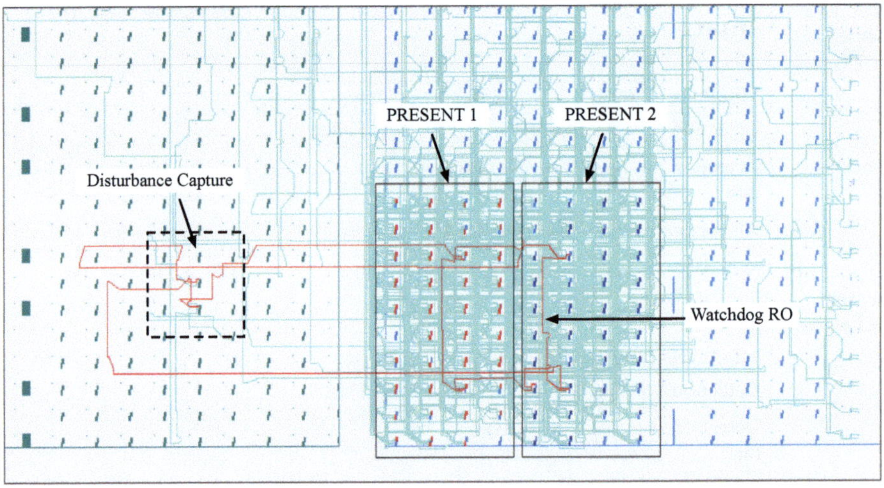

Fig. 11.13 Countermeasure configuration for protecting full PRESENT cipher

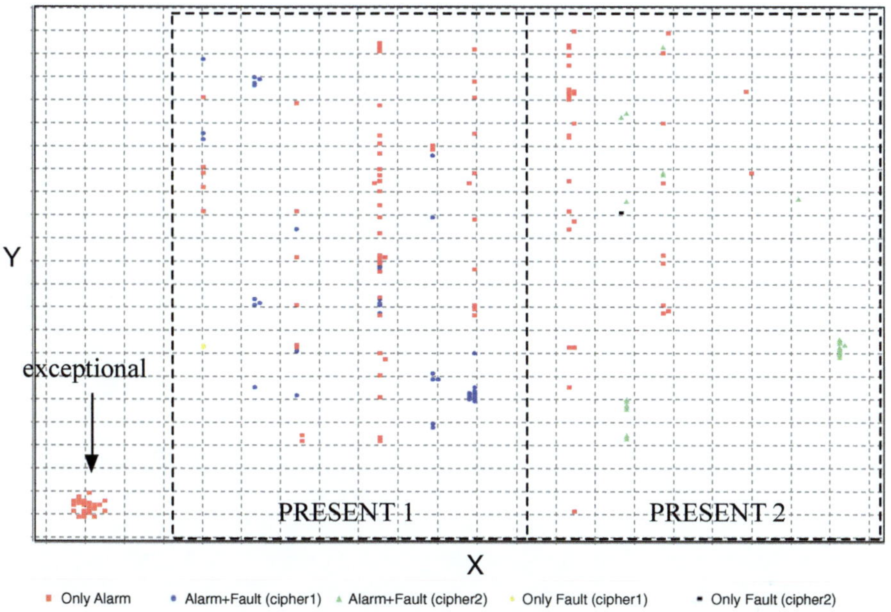

Fig. 11.14 LFI detection experiment on the two full PRESENT-80 ciphers

11.6 Summary

In this IoT era, the low-cost and intelligent embedded devices have facilitated the massively deployed networks comprised by intelligent nodes. However, these low-cost nodes lack the arsenal to resist state of the art attacks like fault injection attacks. In this chapter, we especially focus on the reactive sensor logic at circuit-level against these fault injection threats in consideration of both efficiency and cost. First, a PLL based logic for detecting the on-the-fly fault injection is described, where the instantiated analog PLL module can be found in most modern programmable devices. Next, an all-digital sensor logic with similar function is presented. Both analog and digital sensors are detailed in architecture and sensing mechanisms, and the performances are concretely compared by practical laser injection experiments. The effectiveness of this system is validated on Xilinx 65 nm Virtex-5 FPGA, and the all-digital sensor turned out to be low-cost with high sensitivity. Owing to its pure digital and simple architecture, this system can be easily deployed into any digital/hybrid IC environments, particularly as Internet-of-Things (IoTs) or embedded endpoints of Cyber-Physical System (CPS) with deficient power and hardware resources.

References

1. Agoyan, M., Dutertre, J.M., Mirbaha, A.P., Naccache, D., Ribotta, A.L., Tria, A.: Single-bit DFA using multiple-byte laser fault injection. In: 2010 IEEE International Conference on Technologies for Homeland Security (HST), pp. 113–119. IEEE (2010)
2. Agoyan, M., Dutertre, J.M., Naccache, D., Robisson, B., Tria, A.: When clocks fail: on critical paths and clock faults. Smart Card Research and Advanced Application, pp. 182–193. Springer, Berlin (2010)
3. Alberto, D., Maistri, P., Leveugle, R.: Investigation of electromagnetic fault injection effects on embedded cryptosystems. In: First Workshop on Trustworthy Manufacturing and Utilization of Secure Devices, TRUDEVICE 2013 (2013)
4. Albrecht, M., Cid, C.: Algebraic techniques in differential cryptanalysis. Fast Software Encryption, pp. 193–208. Springer, Berlin (2009)
5. Ali, S., Mukhopadhyay, D.: A differential fault analysis on AES key schedule using single fault. In: 2011 Workshop on Fault Diagnosis and Tolerance in Cryptography, FDTC 2011, Tokyo, Japan, September 29, 2011, pp. 35–42 (2011)
6. Ali, S., Mukhopadhyay, D.: An improved differential fault analysis on AES-256. In: Progress in Cryptology - AFRICACRYPT 2011 - 4th International Conference on Cryptology in Africa, Dakar, Senegal, July 5–7, 2011. Proceedings, pp. 332–347 (2011)
7. Ali, S.S., Mukhopadhyay, D.: Differential fault analysis of twofish. International Conference on Information Security and Cryptology, pp. 10–28. Springer, Berlin (2012)
8. Anderson, R., Kuhn, M.: Tamper resistance-a cautionary note. Proceedings of the Second Usenix Workshop on Electronic Commerce **2**, 1–11 (1996)
9. Anderson, R., Kuhn, M.: Low cost attacks on tamper resistant devices. International Workshop on Security Protocols, pp. 125–136. Springer, Berlin (1997)
10. Anderson, R.J.: Security Engineering: A Guide to Building Dependable Distributed Systems. Wiley, New York (2010)
11. Aumasson, J.P., Henzen, L., Meier, W., Phan, R.C.W.: Sha-3 proposal blake. Submission to NIST (2008)
12. Aweke, Z.B., Yitbarek, S.F., Qiao, R., Das, R., Hicks, M., Oren, Y., Austin, T.M.: ANVIL: software-based protection against next-generation rowhammer attacks. In: Conte, T., Zhou, Y. (eds.) Proceedings of the Twenty-First International Conference on Architectural Support for Programming Languages and Operating Systems, ASPLOS '16, Atlanta, GA, USA, April 2–6, 2016, pp. 743–755. ACM (2016). https://doi.org/10.1145/2872362.2872390
13. Bagheri, N., Ebrahimpour, R., Ghaedi, N.: New differential fault analysis on PRESENT. EURASIP J. Adv. Signal Process. **2013**(1), 1–10 (2013). https://doi.org/10.1186/1687-6180-2013-145

S. Patranabis and D. Mukhopadhyay (eds.), *Fault Tolerant Architectures for Cryptography and Hardware Security*, Computer Architecture and Design Methodologies, https://doi.org/10.1007/978-981-10-1387-4

14. Banik, S., Maitra, S., Sarkar, S.: A differential fault attack on the grain family of stream ciphers. Cryptographic Hardware and Embedded Systems–CHES 2012, pp. 122–139. Springer, Berlin (2012)
15. Banik, S., Maitra, S., Sarkar, S.: A differential fault attack on the grain family under reasonable assumptions. Progress in Cryptology-INDOCRYPT 2012, pp. 191–208. Springer, Berlin (2012)
16. Banik, S., Pandey, S.K., Peyrin, T., Sasaki, Y., Sim, S.M., Todo, Y.: GIFT: a small present towards reaching the limit of lightweight encryption. In: 19th International Conference Cryptographic Hardware and Embedded Systems - CHES 2017, Taipei, Taiwan, September 25–28, 2017 (2017)
17. Bard, G.: Algebraic Cryptanalysis. Springer Science & Business Media (2009)
18. Bard, G.V., Courtois, N.T., Jefferson, C.: Efficient methods for conversion and solution of sparse systems of low-degree multivariate polynomials over GF(2) via SAT-solvers (2007)
19. Barenghi, A., Bertoni, G.M., Breveglieri, L., Pelosi, G.: A fault induction technique based on voltage underfeeding with application to attacks against AES and RSA. J. Syst. Softw. **86**(7), 1864–1878 (2013)
20. Barenghi, A., Breveglieri, L., Koren, I., Naccache, D.: Fault injection attacks on cryptographic devices: theory, practice, and countermeasures. Proc. IEEE **100**(11), 3056–3076 (2012)
21. Barthe, G., Dupressoir, F., Fouque, P.A., Grégoire, B., Zapalowicz, J.C.: Synthesis of fault attacks on cryptographic implementations. In: Proceedings of the 2014 ACM SIGSAC Conference on Computer and Communications Security, pp. 1016–1027. ACM (2014)
22. Battistello, A., Giraud, C.: Fault analysis of infective AES computations. In: Fischer, W., Schmidt, J.M. (eds.) Fault Diagnosis and Tolerance in Cryptography – FDTC 2013, pp. 101–107. IEEE Computer Society (2013)
23. Beaulieu, R., Shors, D., Smith, J., Treatman-Clark, S., Weeks, B., Wingers, L.: The SIMON and SPECK lightweight block ciphers. In: Proceedings of the 52nd Annual Design Automation Conference, San Francisco, CA, USA, June 7–11, 2015, pp. 175:1–175:6 (2015). https://doi.org/10.1145/2744769.2747946
24. Bernstein, D.J.: Chacha, a variant of Salsa20. In: Workshop Record of SASC, vol. 8 (2008)
25. Bernstein, D.J.: The Salsa20 family of stream ciphers. In: New Stream Cipher Designs - The eSTREAM Finalists, pp. 84–97 (2008). https://doi.org/10.1007/978-3-540-68351-3_8
26. Bertacco, V., Alaghi, A., Arthur, W., Tandon, P.: Torturing openSSL
27. Bertoni, G., Breveglieri, L., Koren, I., Maistri, P., Piuri, V.: Error analysis and detection procedures for a hardware implementation of the advanced encryption standard. IEEE Trans. Comput. **52**(4), 492–505 (2003)
28. Berzati, A., Canovas, C., Castagnos, G., Debraize, B., Goubin, L., Gouget, A., Paillier, P., Salgado, S.: Fault analysis of GRAIN-128. In: IEEE International Workshop on Hardware-Oriented Security and Trust, 2009. HOST'09, pp. 7–14. IEEE (2009)
29. Bhattacharya, S., Mukhopadhyay, D.: Curious case of rowhammer: flipping secret exponent bits using timing analysis. In: Cryptographic Hardware and Embedded Systems - CHES 2016 - 18th International Conference, Santa Barbara, CA, USA, August 17–19, 2016, Proceedings, pp. 602–624 (2016)
30. Biham, E., Shamir, A.: Differential Fault Analysis of Secret Key Cryptosystems. In: Jr, B.S.K. (ed.) Advances in Cryptology - CRYPTO 1997. Lecture Notes in Computer Science, vol. 1294, pp. 513–525. Springer, Berlin (1997)
31. Binder, D., Smith, E., Holman, A.: Satellite anomalies from galactic cosmic rays. IEEE Trans. Nucl. Sci. **22**(6), 2675–2680 (1975)
32. Blömer, J., Seifert, J.P.: Fault Based Cryptanalysis of the Advanced Encryption Standard (AES). In: Wright, R.N. (ed.) Financial Cryptography. Lecture Notes in Computer Science, vol. 2742, pp. 162–181. Springer, Berlin (2003)
33. Bogdanov, A., Knudsen, L.R., Leander, G., Paar, C., Poschmann, A., Robshaw, M.J., Seurin, Y., Vikkelsoe, C.: Present: an ultra-lightweight block cipher. In: International Workshop on Cryptographic Hardware and Embedded Systems, pp. 450–466. Springer (2007)

34. Bogdanov, A., Knudsen, L.R., Leander, G., Paar, C., Poschmann, A., Robshaw, M.J.B., Seurin, Y., Vikkelsoe, C.: PRESENT: an ultra-lightweight block cipher. In: Cryptographic Hardware and Embedded Systems - CHES 2007, 9th International Workshop, Vienna, Austria, September 10–13, 2007, Proceedings, pp. 450–466 (2007). https://doi.org/10.1007/978-3-540-74735-2_31
35. Boneh, D., DeMillo, R.A., Lipton, R.J.: On the importance of checking cryptographic protocols for faults (extended abstract). In: Advances in Cryptology - EUROCRYPT '97, International Conference on the Theory and Application of Cryptographic Techniques, Konstanz, Germany, May 11–15, 1997, Proceeding, pp. 37–51 (1997)
36. Boneh, D., Millo, R., Lipton, R.: On the Importance of Checking Cryptographic Protocols for Faults. Advances in Cryptology EUROCRYPT97, pp. 37–51. Springer, Berlin (1997)
37. Bosma, W., Cannon, J., Playoust, C.: The Magma algebra system. I. The user language. J. Symb. Comput. **24**(3–4), 235–265 (1997); Computational Algebra and Number Theory (London, 1993)
38. Bouillaguet, C., Derbez, P., Fouque, P.A.: Automatic search of attacks on round-reduced AES and applications. In: Annual Cryptology Conference, pp. 169–187. Springer (2011)
39. Breier, J., Chen, C.N.: On determining optimal parameters for testing devices against laser fault attacks. In: 2016 International Symposium on Integrated Circuits (ISIC), pp. 1–4. IEEE (2016)
40. Breier, J., He, W.: Multiple fault attack on present with a hardware trojan implementation in FPGA. In: 2015 International Workshop on Secure Internet of Things (SIoT), pp. 58–64 (2015)
41. Breier, J., He, W.: Multiple fault attack on PRESENT with a hardware trojan implementation in FPGA. CoRR (2017). arXiv:1702.08208
42. Buchmann, J.A., Ding, J., Mohamed, M.S.E., Mohamed, W.S.A.E.: Mutantxl: Solving multivariate polynomial equations for cryptanalysis. In: Dagstuhl Seminar Proceedings. Schloss Dagstuhl-Leibniz-Zentrum für Informatik (2009)
43. Buchner, S., Wilson, D., Kang, K., Gill, D., Mazer, J., Raburn, W., Campbell, A., Knudson, A.: Laser simulation of single event upsets. IEEE Trans. Nucl. Sci. **34**(6), 1227–1233 (1987)
44. Buchner, S.P., Miller, F., Pouget, V., McMorrow, D.P.: Pulsed-laser testing for single-event effects investigations. IEEE Trans. Nucl. Sci. **60**(3), 1852–1875 (2013)
45. Chen, C.N., Yen, S.M.: Differential fault analysis on AES key schedule and some countermeasures. In: Goos, G., Hartmanis, J., van Leeuwen, J. (eds.) ACISP 2003, LNCS, vol. 2727, pp. 118–129. Springer (2003)
46. Christofi, M., Chetali, B., Goubin, L., Vigilant, D.: Formal verification of a CRT-RSA implementation against fault attacks. J. Cryptogr. Eng. **3**(3), 157–167 (2013)
47. Cid, C., Murphy, S., Robshaw, M.: Algebraic Aspects of the Advanced Encryption Standard. Springer Science & Business Media (2006)
48. Clavier, C., Feix, B., Gagnerot, G., Roussellet, M.: Passive and active combined attacks on AES combining fault attacks and side channel analysis. In: 2010 Workshop on Fault Diagnosis and Tolerance in Cryptography (FDTC), pp. 10–19. IEEE (2010)
49. Courbon, F., Loubet-Moundi, P., Fournier, J.J., Tria, A.: Adjusting laser injections for fully controlled faults. In: International Workshop on Constructive Side-Channel Analysis and Secure Design, pp. 229–242. Springer (2014)
50. Courbon, F., Loubet-Moundi, P., Fournier, J.J., Tria, A.: Increasing the efficiency of laser fault injections using fast gate level reverse engineering. In: 2014 IEEE International Symposium on Hardware-Oriented Security and Trust (HOST), pp. 60–63. IEEE (2014)
51. Courtois, N., Klimov, A., Patarin, J., Shamir, A.: Efficient algorithms for solving overdefined systems of multivariate polynomial equations. In: International Conference on the Theory and Applications of Cryptographic Techniques, pp. 392–407. Springer (2000)
52. Courtois, N.T.: Algebraic Cryptanalysis: From Plug-and-Pray Experimental Approach to Constructive Optimization. http://www.nicolascourtois.com/papers/fastalg_hisat.pdf
53. Courtois, N.T.: The security of hidden field equations (HFE). In: Cryptographers' Track at the RSA Conference, pp. 266–281. Springer (2001)

54. Courtois, N.T., Bard, G.V.: Algebraic cryptanalysis of the data encryption standard. In: Proceedings of the 11th IMA International Conference on Cryptography and Coding, pp. 152–169. Springer (2007)
55. Courtois, N.T., Jackson, K., Ware, D.: Fault-algebraic attacks on inner rounds of DES. e-Smart'10 Proceedings: The Future of Digital Security Technologies (2010)
56. Courtois, N.T., Pieprzyk, J.: Cryptanalysis of block ciphers with overdefined systems of equations. In: International Conference on the Theory and Application of Cryptology and Information Security, pp. 267–287. Springer (2002)
57. Cox, D., Little, J., O'shea, D.: Ideals, Varieties, and Algorithms, vol. 3. Springer, Berlin (1992)
58. Dassance, F., Venelli, A.: Combined fault and side-channel attacks on the AES key schedule. In: 2012 Workshop on Fault Diagnosis and Tolerance in Cryptography (FDTC), pp. 63–71. IEEE (2012)
59. De Santis, F., Guillen, O.M., Sakic, E., Sigl, G.: Ciphertext-only fault attacks on present. In: Eisenbarth, T., Öztürk, E. (eds.) Lightweight Cryptography for Security and Privacy: Third International Workshop, LightSec 2014, Istanbul, Turkey, September 1–2, 2014, Revised Selected Papers, pp. 85–108. Springer International Publishing, Cham (2015)
60. Decker, W., Greuel, G.M., Pfister, G., Schönemann, H.: SINGULAR 4-1-0 — A computer algebra system for polynomial computations (2016). http://www.singular.uni-kl.de
61. Dehbaoui, A., Dutertre, J.M., Robisson, B., Tria, A.: Electromagnetic transient faults injection on a hardware and a software implementations of AES. In: 2012 Workshop on Fault Diagnosis and Tolerance in Cryptography (FDTC), pp. 7–15. IEEE (2012)
62. Developers, T.S.: SageMath, the Sage Mathematics Software System (Version 7.6) (2017). http://www.sagemath.org
63. Dey, P., Chakraborty, A., Adhikari, A., Mukhopadhyay, D.: Improved practical differential fault analysis of grain-128. In: Proceedings of the 2015 Design, Automation & Test in Europe Conference & Exhibition, DATE 2015, Grenoble, France, March 9–13, 2015, pp. 459–464 (2015). http://dl.acm.org/citation.cfm?id=2755858
64. Dutertre, J.M., Mirbaha, A.P., Naccache, D., Triaz, A.: Reproducible single-byte laser fault injection. In: Ph.D. Research in Microelectronics and Electronics (PRIME), 2010 Conference on, pp. 1–4. IEEE (2010)
65. Fouillat, P., Pouget, V., McMorrow, D., Darracq, F., Buchner, S., Lewis, D.: Fundamentals of the pulsed laser technique for single-event upset testing. Radiation Effects on Embedded Systems, pp. 121–141. Springer, Berlin (2007)
66. Fuhr, T., Jaulmes, E., Lomne, V., Thillard, A.: Fault Attacks on AES with Faulty Ciphertexts Only. 2013 Workshop on Fault Diagnosis and Tolerance in Cryptography(FDTC), pp. 108–118. IEEE (2013)
67. Ghalaty, N., Yuce, B., Taha, M., Schaumont, P.: Differential fault intensity analysis. 2014 Workshop on Fault Diagnosis and Tolerance in Cryptography (FDTC). IEEE (2014)
68. Ghalaty, N.F., Yuce, B., Schaumont, P.: Differential fault intensity analysis on present and LED block ciphers. In: Mangard, S., Poschmann, A.Y. (eds.) Constructive Side-Channel Analysis and Secure Design: 6th International Workshop, COSADE 2015, Berlin, Germany, April 13–14, 2015. Revised Selected Papers, pp. 174–188. Springer International Publishing, Cham (2015)
69. Ghalaty, N.F., Yuce, B., Taha, M., Schaumont, P.: Differential fault intensity analysis. In: 2014 Workshop on Fault Diagnosis and Tolerance in Cryptography (FDTC), pp. 49–58. IEEE (2014)
70. Gierlichs, B., Schmidt, J.M., Tunstall, M.: Infective Computation and Dummy Rounds: Fault Protection for Block Ciphers without Check-before-Output. In: Hevia, A., Neven, G. (eds.) Progress in Cryptology - LATINCRYPT 2012. Lecture Notes in Computer Science, vol. 7533, pp. 305–321. Springer, Berlin (2012)
71. Giraud, C.: DFA on AES. In: Dobbertin, H., Rijmen, V., Sowa, A. (eds.) Advanced Encryption Standard - AES. Lecture Notes in Computer Science, vol. 3373, pp. 27–41. Springer, Berlin (2005)

72. Green, M.A.: Self-consistent optical parameters of intrinsic silicon at 300 K including temperature coefficients. Solar Energy Materials and Solar Cells **92**(11), 1305–1310 (2008)
73. Green, M.A., Keevers, M.J.: Optical properties of intrinsic silicon at 300 K. Progress in Photovoltaics: Research and Applications **3**(3), 189–192 (1995)
74. Gruss, D., Maurice, C., Mangard, S.: Rowhammer.js: a remote software-induced fault attack in javascript. CoRR (2015). arXiv:1507.06955
75. He, W., Breier, J., Bhasin, S.: Cheap and cheerful: a low-cost digital sensor for detecting laser fault injection attacks. In: Security, Privacy, and Applied Cryptography Engineering - 6th International Conference, SPACE 2016, Hyderabad, India, December 14–18, 2016, Proceedings, pp. 27–46 (2016). https://doi.org/10.1007/978-3-319-49445-6_2
76. He, W., Breier, J., Bhasin, S., Miura, N., Nagata, M.: Ring oscillator under laser: Potential of PLL-based countermeasure against laser fault injection. In: 2016 Workshop on Fault Diagnosis and Tolerance in Cryptography, FDTC 2016, Santa Barbara, CA, USA, August 16, 2016, pp. 102–113 (2016). https://doi.org/10.1109/FDTC.2016.13
77. Hell, M., Johansson, T., Maximov, A., Meier, W.: A Stream Cipher Proposal: Grain-128. http://www.ecrypt.eu.org/stream/p3ciphers/grain/Grain128_p3.pdf
78. Huang, R., Yang, H., Chao, M.C., Lin, S.: Alternate hammering test for application-specific drams and an industrial case study. In: Groeneveld, P., Sciuto, D., Hassoun, S. (eds.) The 49th Annual Design Automation Conference 2012, DAC'12, San Francisco, CA, USA, June 3–7, 2012, pp. 1012–1017. ACM (2012). https://doi.org/10.1145/2228360.2228543
79. Hund, R., Willems, C., Holz, T.: Practical timing side channel attacks against kernel space ASLR. In: 2013 IEEE Symposium on Security and Privacy, SP 2013, Berkeley, CA, USA, May 19–22, 2013, pp. 191–205. IEEE Computer Society (2013). https://doi.org/10.1109/SP.2013.23
80. Inci, M.S., Gülmezoglu, B., Irazoqui, G., Eisenbarth, T., Sunar, B.: Cache attacks enable bulk key recovery on the cloud. IACR Cryptology ePrint Archive **2016**, 596 (2016). http://eprint.iacr.org/2016/596
81. Irazoqui, G., Eisenbarth, T., Sunar, B.: Systematic reverse engineering of cache slice selection in intel processors. In: 2015 Euromicro Conference on Digital System Design, DSD 2015, Madeira, Portugal, August 26–28, 2015, pp. 629–636. IEEE Computer Society (2015). https://doi.org/10.1109/DSD.2015.56
82. JEDEC: Standard No. 79-3F. DDR3 SDRAM Specification (2012)
83. JEDEC Solid State Technology Association: Low Power Double Data Rate 4 (LPDDR4) (2015)
84. Jiang, W., Wen, L., Jiang, K., Zhang, X., Pan, X., Zhou, K.: System-level design to detect fault injection attacks on embedded real-time applications. JETC **13**(2), 22:1–22:18 (2016). https://doi.org/10.1145/2967611
85. Jovanovic, P., Kreuzer, M., Polian, I.: An algebraic fault attack on the led block cipher. IACR Cryptology ePrint Archive **2012**, 400 (2012)
86. Joye, M., Manet, P., Rigaud, J.B.: Strengthening hardware aes implementations against fault attacks. IET Information Security **1**(3), 106–110 (2007)
87. Junod, P., Vaudenay, S.: Perfect diffusion primitives for block ciphers. Selected Areas in Cryptography, pp. 84–99. Springer, Berlin (2005)
88. Karmakar, S., Chowdhury, D.R.: Fault analysis of Grain-128 by targeting NFSR. In: Progress in Cryptology–AFRICACRYPT 2011, pp. 298–315. Springer (2011)
89. Karpovsky, M., Kulikowski, K.J., Taubin, A.: Robust protection against fault-injection attacks on smart cards implementing the advanced encryption standard. In: 2004 International Conference on Dependable Systems and Networks, pp. 93–101. IEEE (2004)
90. Karri, R., Wu, K., Mishra, P., Kim, Y.: Concurrent error detection of fault-based side-channel cryptanalysis of 128-bit symmetric block ciphers. In: DAC, pp. 579–585 (2001)
91. Karri, R., Wu, K., Mishra, P., Kim, Y.: Fault-based side-channel cryptanalysis tolerant Rijndael symmetric block cipher architecture. In: DFT, pp. 427–435 (2001)
92. Karri, R., Wu, K., Mishra, P., Kim, Y.: Concurrent error detection schemes for fault-based side-channel cryptanalysis of symmetric block ciphers. IEEE Trans. Comput.-Aided Des. Integr. Circuits Syst. **21**(12), 1509–1517 (2002)

93. Kermani, M.M., Reyhani-Masoleh, A.: Parity prediction of S-box for AES. In: CCECE, pp. 2357–2360 (2006)
94. Kermani, M.M., Reyhani-Masoleh, A.: A Low-cost S-box for the advanced encryption standard using normal basis. In: EIT, pp. 52–55 (2009)
95. Khanna, P., Rebeiro, C., Hazra, A.: XFC: A framework for eXploitable fault characterization in block ciphers. In: DAC 2017. IEEE (2017) (To Appear)
96. Kim, C.H.: Differential fault analysis against AES-192 and AES-256 with minimal faults. In: 2010 Workshop on Fault Diagnosis and Tolerance in Cryptography (FDTC), pp. 3–9. IEEE (2010)
97. Kim, C.H., Quisquater, J.: New differential fault analysis on AES key schedule: two faults are enough. In: Smart Card Research and Advanced Applications, 8th IFIP WG 8.8/11.2 International Conference, CARDIS 2008, London, UK, September 8–11, 2008. Proceedings, pp. 48–60 (2008)
98. Kim, D., Nair, P.J., Qureshi, M.K.: Architectural support for mitigating row hammering in DRAM memories. Comput. Archit. Lett. **14**(1), 9–12 (2015). https://doi.org/10.1109/LCA.2014.2332177
99. Kim, H.: Differential Fault Analysis against AES-192 and AES-256 with Minimal Faults. 2010 Workshop on Fault Diagnosis and Tolerance in Cryptography (FDTC), pp. 3–9. IEEE (2010)
100. Kim, H.: Improved differential fault analysis on AES key schedule. IEEE Trans. Inf. Forensics Secur. **7**(1), 41–50 (2012)
101. Kim, Y., Daly, R., Kim, J., Fallin, C., Lee, J., Lee, D., Wilkerson, C., Lai, K., Mutlu, O.: Flipping bits in memory without accessing them: an experimental study of DRAM disturbance errors. In: ACM/IEEE 41st International Symposium on Computer Architecture, ISCA 2014, Minneapolis, MN, USA, June 14–18, 2014, pp. 361–372. IEEE Computer Society (2014). https://doi.org/10.1109/ISCA.2014.6853210
102. Kipnis, A., Shamir, A.: Cryptanalysis of the HFE public key cryptosystem by relinearization. In: Annual International Cryptology Conference, pp. 19–30. Springer (1999)
103. Knudsen, L.R., Miolane, C.V.: Counting equations in algebraic attacks on block ciphers. Int. J. Inf. Secur. **9**(2), 127–135 (2010)
104. Kocar, O.: Hardwaresicherheit von mikrochips in chipkarten. Datenschutz und Datensicherheit **20**(7), 421–424 (1996)
105. Kömmerling, O., Kuhn, M.G.: Design principles for tamper-resistant smartcard processors. Smartcard **99**, 9–20 (1999)
106. Lewis, D., Pouget, V., Beaudoin, F., Perdu, P., Lapuyade, H., Fouillat, P., Touboul, A.: Backside laser testing of ICs for set sensitivity evaluation. IEEE Trans. Nucl. Sci. **48**(6), 2193–2201 (2001)
107. Li, W., Gu, D., Wang, Y., Li, J., Liu, Z.: An extension of differential fault analysis on AES. In: Third International Conference on Network and System Security, pp. 443–446. NSS (2009)
108. Li, Y., Endo, S., Debande, N., Homma, N., Aoki, T., Le, T.H., Danger, J.L., Ohta, K., Sakiyama, K.: Exploring the relations between fault sensitivity and power consumption. In: Constructive Side-Channel Analysis and Secure Design, pp. 137–153. Springer (2013)
109. Li, Y., Hayashi, Y.i., Matsubara, A., Homma, N., Aoki, T., Ohta, K., Sakiyama, K.: Yet another fault-based leakage in non-uniform faulty ciphertexts. In: Foundations and Practice of Security, pp. 272–287. Springer (2014)
110. Li, Y., Sakiyama, K., Gomisawa, S., Fukunaga, T., Takahashi, J., Ohta, K.: Fault sensitivity analysis. In: Cryptographic Hardware and Embedded Systems, CHES 2010, 12th International Workshop, Santa Barbara, CA, USA, August 17–20, 2010. Proceedings, pp. 320–334 (2010). https://doi.org/10.1007/978-3-642-15031-9_22
111. Li, Y., Sakiyama, K., Gomisawa, S., Fukunaga, T., Takahashi, J., Ohta, K.: Fault Sensitivity Analysis. In: Cryptographic Hardware and Embedded Systems, CHES 2010, pp. 320–334. Springer (2010)
112. Li, Y., Sakiyama, K., Gomisawa, S., Fukunaga, T., Takahashi, J., Ohta, K.: Fault sensitivity analysis. Cryptographic Hardware and Embedded Systems, CHES 2010, pp. 320–334. Springer (2010)

113. Liu, F., Yarom, Y., Ge, Q., Heiser, G., Lee, R.B.: Last-level cache side-channel attacks are practical. In: 2015 IEEE Symposium on Security and Privacy, SP 2015, San Jose, CA, USA, May 17–21, 2015, pp. 605–622. IEEE Computer Society (2015). https://doi.org/10.1109/SP.2015.43
114. Lomné, V., Roche, T., Thillard, A.: On the need of randomness in fault attack countermeasures - application to AES. In: Bertoni, G., Gierlichs, B. (eds.) Fault Diagnosis and Tolerance in Cryptography – FDTC 2012, pp. 85–94. IEEE Computer Society (2012)
115. Maher, D.P.: Fault induction attacks, tamper resistance, and hostile reverse engineering in perspective. In: International Conference on Financial Cryptography, pp. 109–121. Springer (1997)
116. Maistri, P., Leveugle, R.: Double-data-rate computation as a countermeasure against fault analysis. IEEE Trans. Comput. **57**(11), 1528–1539 (2008)
117. Mansouri, S.S., Dubrova, E.: An improved hardware implementation of the grain stream cipher. In: 2010 13th Euromicro Conference on Digital System Design: Architectures, Methods and Tools (DSD), pp. 433–440. IEEE (2010)
118. Seaborn, M., Dullien, T.: Exploiting the DRAM rowhammer bug to gain kernel privileges (2015). http://googleprojectzero.blogspot.in/2015/03/exploiting-dram-rowhammer-bug-to-gain.html
119. Seaborn, M., Dullien, T.: Test DRAM for bit flips caused by the rowhammer problem (2015). https://github.com/google/rowhammer-test
120. Maurice, C., Scouarnec, N.L., Neumann, C., Heen, O., Francillon, A.: Reverse engineering intel last-level cache complex addressing using performance counters. In: Research in Attacks, Intrusions, and Defenses - 18th International Symposium, RAID 2015, Kyoto, Japan, November 2–4, 2015, Proceedings, pp. 48–65 (2015)
121. Micron Inc.: DDR4 SDRAM MT40A2G4, MT40A1G8, MT40A512M16 Data sheet, 2015 (2015)
122. Miura, N., Najm, Z., He, W., Bhasin, S., Ngo, X.T., Nagata, M., Danger, J.: PLL to the rescue: a novel EM fault countermeasure. In: Proceedings of the 53rd Annual Design Automation Conference, DAC 2016, Austin, TX, USA, June 5–9, 2016, pp. 90:1–90:6 (2016). https://doi.org/10.1145/2897937.2898065
123. Moradi, A., Mischke, O., Paar, C., Li, Y., Ohta, K., Sakiyama, K.: On the power of fault sensitivity analysis and collision side-channel attacks in a combined setting. Cryptographic Hardware and Embedded Systems–CHES 2011, p. 292
124. Morioka, S., Satoh, A.: An optimized s-box circuit architecture for low power AES design. In: Jr., B.S.K., Koç, Ç.K., Paar, C (eds.) Cryptographic Hardware and Embedded Systems - CHES 2002, 4th International Workshop, Redwood Shores, CA, USA, August 13–15, 2002, Revised Papers, Lecture Notes in Computer Science, vol. 2523, pp. 172–186. Springer (2002). https://doi.org/10.1007/3-540-36400-5_14
125. Moro, N., Dehbaoui, A., Heydemann, K., Robisson, B., Encrenaz, E.: Electromagnetic fault injection: towards a fault model on a 32-bit microcontroller. In: 2013 Workshop on Fault Diagnosis and Tolerance in Cryptography (FDTC), pp. 77–88. IEEE (2013)
126. Moro, N., Heydemann, K., Encrenaz, E., Robisson, B.: Formal verification of a software countermeasure against instruction skip attacks. J. Cryptogr. Eng. **4**(3), 145–156 (2014)
127. Mozaffari-Kermani, M., Reyhani-Masoleh, A.: Parity-based fault detection architecture of S-box for advanced encryption standard. In: DFT, pp. 572–580 (2006)
128. Mozaffari-Kermani, M., Reyhani-Masoleh, A.: A lightweight concurrent error detection scheme for the AES S-boxes using normal basis. In: Proceedings of CHES, pp. 113–129 (2008)
129. Mozaffari-Kermani, M., Reyhani-Masoleh, A.: Concurrent structure-independent fault detection schemes for the advanced encryption standard. IEEE Trans. Comput. **59**(5), 608–622 (2010)
130. Mozaffari-Kermani, M., Reyhani-Masoleh, A.: A lightweight high-performance fault detection scheme for the advanced encryption standard using composite field. IEEE Trans. VLSI Syst. **19**(1), 85–91 (2011)

131. Mukhopadhyay, D.: An improved fault based attack of the advanced encryption standard. In: Preneel, B. (ed.) Progress in Cryptology - AFRICACRYPT 2009. Lecture Notes in Computer Science, vol. 5580, pp. 421–434. Springer, Berlin (2009)
132. Mukhopadhyay, D., Chakraborty, R.S.: Hardware Security - Design, Threats, and Safeguards. CRC Press, Boca Raton (2014)
133. Murali, S., Srikanth, N.: Acid decapsulation of epoxy molded IC packages with copper wire bonds. IEEE Trans. Electron. Packag. Manuf. **29**(3), 179–183 (2006)
134. Naccache, D., M'Raihi, D.: Cryptographic smart cards. IEEE Micro **16**(3), 14–24 (1996)
135. Natale, G.D., Flottes, M.L., Rouzeyre, B.: A novel parity bit scheme for S-box in AES Circuits. In: DDECS, pp. 1–5 (2007)
136. Natale, G.D., Flottes, M.L., Rouzeyre, B.: On-line self-test of AES hardware implementation. WDSN (2007)
137. Nyberg, K.: Differentially uniform mappings for cryptography. In: EUROCRYPT, pp. 55–64 (1993)
138. Oren, Y., Kirschbaum, M., Popp, T., Wool, A.: Algebraic side-channel analysis in the presence of errors. In: International Workshop on Cryptographic Hardware and Embedded Systems, pp. 428–442. Springer, Berlin (2010)
139. Osvik, D.A., Shamir, A., Tromer, E.: Cache attacks and countermeasures: the case of AES. In: Pointcheval, D. (ed.) CT-RSA. Lecture Notes in Computer Science, vol. 3860, pp. 1–20. Springer, Berlin (2006)
140. Patarin, J.: Cryptanalysis of the Matsumoto and Imai public key scheme of Eurocrypt'88. In: Annual International Cryptology Conference, pp. 248–261. Springer (1995)
141. Patranabis, S., Chakraborty, A., Mukhopadhyay, D., Chakrabarti, P.P.: Fault space transformation: a generic approach to counter differential fault analysis and differential fault intensity analysis on AES-like block ciphers. IEEE Trans. Inf. Forensics Secur. **12**(5), 1092–1102 (2017)
142. Peacham, D., Thomas, B.: A DFA attack against the AES key schedule. SiVenture White Paper 001, 26 October (2006)
143. Pessl, P., Gruss, D., Maurice, C., Mangard, S.: Reverse engineering intel DRAM addressing and exploitation. CoRR (2015). arXiv:1511.08756
144. Peterson, I.: Chinks in digital armor: exploiting faults to break smart-card cryptosystems. Sci. News **151**(5), 78–79 (1997)
145. Petit, C., Kosters, M., Messeng, A.: Algebraic approaches for the elliptic curve discrete logarithm problem over prime fields. In: IACR International Workshop on Public Key Cryptography, pp. 3–18. Springer (2016)
146. Piret, G., Quisquater, J.J.: A differential fault attack technique against SPN structures, with Application to the AES and KHAZAD. Cryptographic Hardware and Embedded Systems, CHES 2003, pp. 77–88. Springer (2003)
147. Piret, G., Quisquater, J.J.: A Differential fault attack technique against SPN structures, with Application to the AES and KHAZAD. In: Walter, C.D., KoÇ, Ç.K., Paar, C (eds.) Cryptographic Hardware and Embedded Systems - CHES 2003, Lecture Notes in Computer Science, vol. 2779, pp. 77–88. Springer (2003)
148. Qiao, R., Seaborn, M.: A new approach for rowhammer attacks. In: HOST, 2016 (2016). http://seclab.cs.sunysb.edu/seclab/pubs/host16.pdf
149. Quisquater, J.J., Samyde, D.: ElectroMagnetic Analysis (EMA): Measures and countermeasures for smard cards. In: Attali, I., Jensen, T.P. (eds.) E-smart, LNCS, vol. 1240, pp. 200–210. Springer, Berlin (2001). ISSN 0302-9743
150. Quisquater, J.J., Samyde, D.: Eddy current for magnetic analysis with active sensor. In: Proceedings of Esmart, vol. 2002 (2002)
151. Rajendran, J., Borad, H., Mantravadi, S., Karri, R.: SLICED: Slide-based concurrent error detection technique for symmetric block cipher. In: HOST, pp. 70–75 (2010)
152. Rauzy, P., Guilley, S.: A formal proof of countermeasures against fault injection attacks on CRT-RSA. J. Cryptogr. Eng. **4**(3), 173–185 (2014)

153. Renauld, M., Standaert, F.X.: Algebraic side-channel attacks. International Conference on Information Security and Cryptology, pp. 393–410. Springer, Berlin (2009)
154. Rivain, M.: Differential fault analysis on DES middle rounds. Cryptographic Hardware and Embedded Systems-CHES, pp. 457–469. Springer, Berlin (2009)
155. Robisson, B., Manet, P.: Differential behavioral analysis. In: Cryptographic Hardware and Embedded Systems - CHES 2007, 9th International Workshop, Vienna, Austria, September 10–13, 2007, Proceedings, pp. 413–426 (2007). https://doi.org/10.1007/978-3-540-74735-2_28
156. Roche, T., Lomné, V., Khalfallah, K.: Combined fault and side-channel attack on protected implementations of AES. In: Smart Card Research and Advanced Applications, pp. 65–83. Springer, Berlin (2011)
157. Rohani, N., Noferesti, Z., Mohajeri, J., Aref, M.R.: Guess and determine attack on Trivium family. In: 2010 IEEE/IFIP 8th International Conference on Embedded and Ubiquitous Computing (EUC), pp. 785–790. IEEE (2010)
158. Roscian, C., Dutertre, J.M., Tria, A.: Frontside laser fault injection on cryptosystems-application to the AES'last round. In: 2013 IEEE International Symposium on Hardware-Oriented Security and Trust (HOST), pp. 119–124. IEEE (2013)
159. Roscian, C., Sarafianos, A., Dutertre, J.M., Tria, A.: Fault model analysis of laser-induced faults in SRAM memory cells. In: 2013 Workshop on Fault Diagnosis and Tolerance in Cryptography (FDTC), pp. 89–98. IEEE (2013)
160. Saritaş, M., McKell, H.D.: Absorption coefficient of si in the wavelength region between 0.80–1.16 μm. J. Appl. Phy. **61**(10), 4923–4925 (1987)
161. Sarkar, S., Banik, S., Maitra, S.: Differential fault attack against grain family with very few faults and minimal assumptions. IACR Cryptol. ePrint Arch. **2013**, 494 (2013)
162. Satoh, A., Sugawara, T., Homma, N., Aoki, T.: High-performance concurrent error detection scheme for AES hardware. In: Cryptographic Hardware and Embedded Systems–CHES 2008, pp. 100–112. Springer (2008)
163. Schneier, B., Kelsey, J., Whiting, D., Wagner, D., Hall, C., Ferguson, N.: Twofish: a 128-bit block cipher. NIST AES Proposal **15** (1998)
164. Selmane, N., Bhasin, S., Guilley, S., Graba, T., Danger, J.L.: WDDL is protected against setup time violation attacks. In Fault Diagnosis and Tolerance in Cryptography (FDTC), 2009 Workshop on, September 2009, 73–83, IEEE (2009). https://doi.org/10.1109/FDTC.2009.40
165. Shannon, C.E.: Communication theory of secrecy systems. Bell Labs Tech. J. **28**(4), 656–715 (1949)
166. Skorobogatov, S.: Optically enhanced position-locked power analysis. In: Cryptographic Hardware and Embedded Systems-CHES 2006, pp. 61–75. Springer (2006)
167. Skorobogatov, S.P., Anderson, R.J., et al.: Optical fault induction attacks. In: CHES, vol. 2523, pp. 2–12. Springer (2002)
168. Soos, M.: Cryptominisat–a sat solver for cryptographic problems (2009), http://www.msoos.org/cryptominisat4
169. Takahashi, J., Fukunaga, T.: Differential fault analysis on AES with 192 and 256-Bit keys. Cryptology ePrint Archive, Report 2010/023 (2010). http://eprint.iacr.org/
170. Takahashi, J., Fukunaga, T., Yamakoshi, K.: DFA mechanism on the AES key schedule. In: FDTC, pp. 62–74 (2007)
171. Trichina, E., Korkikyan, R.: Multi fault laser attacks on protected CRT-RSA. In: 2010 Workshop on Fault Diagnosis and Tolerance in Cryptography (FDTC), pp. 75–86. IEEE (2010)
172. Tunstall, M., Mukhopadhyay, D., Ali, S.: Differential fault analysis of the advanced encryption standard using a single fault. Information Security Theory and Practice. Security and Privacy of Mobile Devices in Wireless Communication, pp. 224–233. Springer, Berlin (2011)
173. Tupsamudre, H., Bisht, S., Mukhopadhyay, D.: Destroying fault invariant with randomization. In: Cryptographic Hardware and Embedded Systems–CHES 2014, pp. 93–111. Springer (2014)

174. van der Veen, V., Fratantonio, Y., Lindorfer, M., Gruss, D., Maurice, C., Vigna, G., Bos, H., Razavi, K., Giuffrida, C.: Drammer: Deterministic rowhammer attacks on mobile platforms. In: Proceedings of the 2016 ACM SIGSAC Conference on Computer and Communications Security, Vienna, Austria, October 24–28, 2016, pp. 1675–1689 (2016)
175. Verbauwhede, I., Tiri, K.J.: Wave dynamic differential logic. U.S. Patent 8,947,123 (2015)
176. Wang, G., Wang, S.: Differential fault analysis on PRESENT key schedule. In: 2010 International Conference on Computational Intelligence and Security (CIS), pp. 362–366 (2010). https://doi.org/10.1109/CIS.2010.84
177. Wikipedia: Rowhammer wikipedia page (2016). https://en.wikipedia.org/wiki/Row-hammer
178. Wu, K., Karri, R., Kuznetsov, G., Goessel, M.: Low cost concurrent error detection for the advanced encryption standard. In: Test Conference, 2004. Proceedings. ITC 2004. International, pp. 1242–1248. IEEE (2004)
179. Yarom, Y., Falkner, K.: FLUSH+RELOAD: a high resolution, low noise, L3 cache side-channel attack. In: Fu, K., Jung, J. (eds.) Proceedings of the 23rd USENIX Security Symposium, San Diego, CA, USA, August 20–22, 2014., pp. 719–732. USENIX Association (2014). https://www.usenix.org/conference/usenixsecurity14/technical-sessions/presentation/yarom
180. Zhang, F., Guo, S., Zhao, X., Wang, T., Yang, J., Standaert, F.X., Gu, D.: A framework for the analysis and evaluation of algebraic fault attacks on lightweight block ciphers. IEEE Trans. Inf. Forensics Secur. **11**(5), 1039–1054 (2016)
181. Zhang, F., Zhao, X., Guo, S., Wang, T., Shi, Z.: Improved algebraic fault analysis: a case study on piccolo and applications to other lightweight block ciphers. In: International Workshop on Constructive Side-Channel Analysis and Secure Design, pp. 62–79. Springer (2013)
182. Zhang, W., Bao, Z., Lin, D., Rijmen, V., Yang, B., Verbauwhede, I.: RECTANGLE: a bit-slice lightweight block cipher suitable for multiple platforms. Sci. China Inf. Sci. **58**(12), 1–15 (2015). https://doi.org/10.1007/s11432-015-5459-7
183. Zhao, X., Guo, S., Wang, T., Zhang, F., Shi, Z.: Fault-propagate pattern based DFA on PRESENT and PRINT cipher. Wuhan Univ. J. Nat. Sci. **17**(6), 485–493 (2012). https://doi.org/10.1007/s11859-012-0875-7
184. Zhao, X., Zhang, F., Guo, S., Wang, T., Shi, Z., Liu, H., Ji, K.: MDASCA: an enhanced algebraic side-channel attack for error tolerance and new leakage model exploitation. In: International Workshop on Constructive Side-Channel Analysis and Secure Design, pp. 231–248. Springer (2012)
185. Zhao, X.J., Guo, S., Zhang, F., Wang, T., Shi, Z., Ji, K.: Algebraic differential fault attacks on led using a single fault injection. IACR Cryptol. ePrint Arch. **2012**, 347 (2012)
186. Zhao, X.J., Wang, T., Guo, S.: Fault-propagation pattern based DFA on spn structure block ciphers using bitwise permutation, with application to present and printcipher. IACR Cryptol. ePrint Arch. **2011**, 86 (2011)

MIX
Papier aus verantwortungsvollen Quellen
Paper from responsible sources
FSC® C105338

If you have any concerns about our products,
you can contact us on
ProductSafety@springernature.com

In case Publisher is established outside the EU,
the EU authorized representative is:
Springer Nature Customer Service Center GmbH
Europaplatz 3, 69115 Heidelberg, Germany

Printed by Libri Plureos GmbH
in Hamburg, Germany